Page 1
Detail from AJ Fisher's 1909 patent for a washing machine.

Pages 2 and 3
A visit from HC Booth's vacuum cleaner.

Photographic acknowledgments

All illustrations in this book have been taken from out-of-copyright material held in the British Library's collections, with the exception of those which carry a credit line in the accompanying caption. Patents are numbered and dated for ease of reference.

OXFORD UNIVERSITY PRESS

Oxford New York Toronto
Delhi Bombay Calcutta Madras Karachi
Kuala Lumpur Singapore Hong Kong Tokyo
Nairobi Dar es Salaam Cape Town
Melbourne Auckland Madrid
and associated companies in
Berlin Ibadan

Library of Congress Cataloging in Publication Data

Weaver, Rebecca.
    Machines in the home / Rebecca Weaver, Rodney Dale.
    p.64, 24.6 x 18.9 cm. — (Discoveries and inventions)
    Includes bibliographical references (p. 63) and index.
    Summary: Discusses the development of domestic technology—heating and lighting, cooking, laundry, the modern bathroom, carpet sweeping, and assorted gadgetry.
    ISBN 0-19-520965-6 (acid-free paper)
    1. Household appliances—History—Juvenile literature.
    2. Technology—History—Juvenile literature.
    [1.Household appliances—History. 2. Technology—History.]
    I. Dale, Rodney, 1933- . II. Title. III. Series.
TX298.W43 1992
683'.8—dc20   92-21662
                                                    CIP
                                                    AC

ISBN 0-19-520969-9 (paperback)
ISBN 0-19-520965-6 (hardback)
Printing (last digit) 9 8 7 6 5 4 3 2 1

Designed and set on Ventura in Palatino by Roger Davies
Printed in Singapore

Contents

"STUART" INDEPENDENT BATH

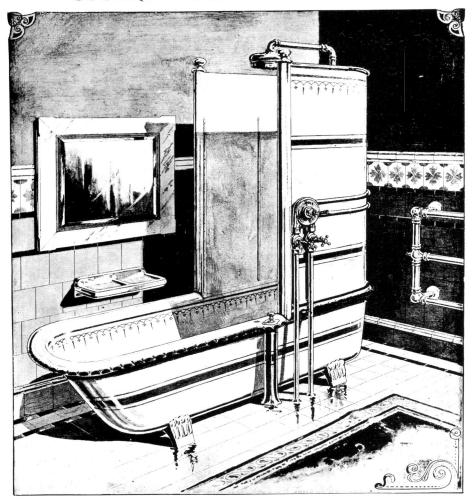

No. 05 GEORGE FARMILOE & SONS' "STUART" INDEPENDENT BATH
CONSISTING OF

6′ 0″ × 2′ 0″ Parallel Cast-iron Bath, Best Metallic Enamelled Interior and Enamelled Decorated Exterior, fitted with 1″ "Hot" and "Cold" Taps, and Registered Mixing and Dividing Arrangement for Shower, Spray and Wave Supply, Copper Supply Pipes and Lift-up Waste as shown. Spray Enclosure made of Plate Zinc

	£	s.	d.
Spray Enclosure made of Plate Zinc	28	5	0
No. 06 If **Bath** is 6′ 0″ × 2′ 3″	29	7	6

Made also left hand.

EXTRAS—	£	s.	d.
Nickel-plated Fittings, Shower and Hot and Cold Pipes	0	11	0
Copper Spray Enclosure instead of Zinc	8	3	6
Uprising Spray or Uprising Douche	1	18	0
If Reclining Bath is Vitreous Enamelled inside	1	8	6
,, ,, ,, White Porcelain Enamelled inside	2	5	0
Towel Rail 3′ 0″ × 2′ 6″	5	14	6
Whiteware Sponge and Soap Tray	0	10	0
Bevelled Edge Mirror (interior 2′ 6″ × 1′ 8″) with Polished Veined Marble frame (3′ 2″ × 2′ 4″ outside) and Clips	7	1	6

Introduction

The Great Exhibition of 1851 drew, as Queen Victoria herself said, 'but one voice of astonishment and admiration'. It confirmed Britain's place as world leader in technology and seemed to usher in a new era of industrial confidence and expansion. Versions of nearly all the appliances illustrated in this book were shown at the Exhibition; the fact that many subsequently took so long to become an accepted part of our lives is as much a result of conservatism and a struggle to interest the market as of the inability of science to meet the challenge of technological concepts and demands. For example, the Royal Commission responsible for setting up the Exhibition refused to allow any appliance to be exhibited which was 'in practical operation through the agents of gas', thereby limiting the impact of a whole range of useful devices.

By the time of the Exhibition, industrialization had reached a point at which it was possible to fulfil the demand for domestic appliances for a growing market. Steam-powered machines were increasingly common in factories for the production of parts, both new and replacement. Yet the spirit of enthusiastic creativity characteristic of the 100 or so years prior to the Exhibition was being contained and tamed by a sobriety, sedateness and smugness we now recognize as the Victorian spirit. Even though growing numbers of the working classes were, in the last quarter of the 19th century, reaping the benefits of industry and trade, there was a fine sense of discrimination as to what elements of modernity were adopted. Perhaps it was merely a combination of conservatism and pragmatism that recognized gas lighting but not gas cooking; a bath and a water closet* (if you had the water supplies) but not a washing machine.

The history of domestic water supply and disposal appears to be equally haphazard on both sides of the Atlantic. The initiative to pump and supply was earlier more evident in the United States – though it appears that towns were often more eager to supply than to dispose and the systems quickly became overloaded as demand grew. Even Catherine E Beecher (sister of Harriet Beecher Stowe of Uncle Tom's Cabin fame) advocated earth closets as a more practical alternative to the expense of plumbing installations 'and the inevitable disorders of water pipes in a house'.

There was a strong business lobby in North America which spearheaded demand. Cities which wanted to compete for commercial trade had to gain a reputation for cleanliness. But by the time the Depression halted economic expansion, Zanesville, OH, for example, could claim 91 per cent of its homes receiving running water, with 61 per cent boasting complete plumbing systems.

The problem was slightly different in Britain where, from the 1870s, new houses were automatically plumbed. Existing dwellings in older cities, however, posed greater obstacles so that, although almost every street in Oxford, for example, had a water supply by 1912, taps were often out in the open and had to be shared. In 1934 London, working-class tenements were not individually supplied and the water tap could be on a landing, down three flights of stairs, or across a yard. The 1951 census reports that only 59 per cent of households in England and Wales could boast a kitchen sink, a fixed bath, and a water closet.

Gas had been used to light urban streets – and subsequently factories, public buildings and shops – since early in the 19th century. It was easy to manufacture and by the time of the 1851 Exhibition more than 900

---

* In England, the 'closet' – itself euphemistic – became the lavatory (the place where one washes) and then the toilet (the act of washing). Closets were of two types. In the earth closet, a shower of earth covers the newly-deposited contents; the bucket therefore has to be emptied from time to time. In the water closet, the goods were washed away with water – into a sewage system if you were lucky. We want to distinguish between the use of earth and water, and so adopt the terms earth closet and water closet.

towns and cities in Britain had their own supplies. In the United States too, by the outbreak of the Civil War, making and distributing gas was a major enterprise.

It would therefore have been relatively simple to extend the supply, yet little was done until the clean, bright light of the electric lamp prompted the gas industry to meet the competition and initiate a programme of public awareness and confidence-raising.

In Britain, appliances were leased for would-be consumers to experiment with, and in the 1890s penny-in-the-slot meters became a standard way of paying for domestic consumption, enabling customers to pay for the gas they used as they used it. In the United States, the trade association of gas companies, the American Gas Association, decided that co-operation was the way forward and subsidized work on the improvement of gas stoves. In 1899, 75 per cent of the gas produced nationwide was used for illumination. By 1921, only 21 per cent was used for illumination, and 54 per cent for domestic fuel. The rest was consumed by industry.

Generating and distributing electricity was more complex. The whole question of whether or not it ought to be made available to the mass of population became the discussion topic of the 1890s. Even Rookes Evelyn Bell Crompton (1845–1940), pioneer of the electricity supply industry, observed that 'the electric current was too expensive to become general'. It certainly required great capital investment and extraordinary amounts of co-ordination between different companies. In Britain, little real progress was made until the Electricity (Supply) Act of 1926. Apart from setting up the national grid, companies were encouraged to woo customers – again by leasing appliances – and the hitherto very high price of electricity fell considerably. The proportion of households with electrical services jumped from 18 per cent in 1926 to 65 per cent in 1938 and 86 per cent in 1949. Having said that, apart from lighting, most of that supply was consumed by irons

Top
Then, as now, people complained of refuse and offal filling the moat and destroying the fish.

Below
The luxury of a hot bath; Queen Elizabeth I took one 'once a year, whether she needed it or not'!

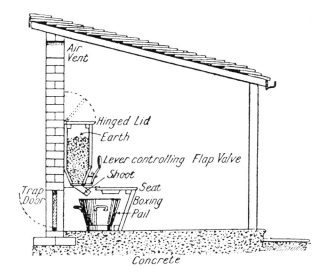

The workings of an earth closet.

and space heating. Despite the not inconsiderable efforts of the pressure group, the Electrical Association for Women, large-scale purchases of electrical appliances had to wait until the affluent decades of the later 20th century.

Standardizing generation and supply was achieved much, much earlier in the United States. So few companies held crucial patents for the manufacture of electrical goods that it was a relatively simple matter for them to agree on standards. Federal governments encouraged cross-licensing (patents licensed to more than one company) as a technique for avoiding charges of monopolistic control. By 1910, the entire country had become standardized. Setting up to mass produce electrical goods became a worthwhile investment; the prices of such goods fell, and stimulated demand. In 1907, eight per cent of homes were wired; by 1940 80 per cent were. After lighting, the main demand was exactly the same as in Britain – for irons.

It is often said that, as household appliances increased, so the servant population declined. In fact, most important developments took place when servants were numerous. But, as we have suggested above, it was not the presence of cheap labour alone that prevented these developments from becoming instant successes. On the other hand, those of the aspiring middle class were encouraged by popular and trade journals to resist the inclination to flaunt their

status by employing a maid and instead to embrace modernity and invest in an appliance or two.

The American middle classes were encouraged to take a very expedient attitude to servants. 'In England the class who go to service are a class, and service is a profession . . . In America, domestic service is a stepping stone to something higher.' This pronouncement came from Catherine E Beecher in 1869; she adopted a most prescriptive response to household appliances. For her, they were the most effective way of minimizing the effort expended on housework. She believed that appliances could, properly used or even pooled, free women to spend their time more fruitfully. She advocated, for example, one washing machine for every 12 households.

Less political in her approach, and clearly rooted in the 'servants-as-a-class' culture, was Britain's Isabella Mary Beeton. Her *Household Management* was first published in parts in 1859–61, and fast became the most influential manual of its day. Even after her death, updated editions were instrumental in giving an accepted, though sometimes belatedly cautious, blessing to new household devices and appliances. In towns particularly, alternative employment opportunities in offices, factories, shops, nursing and teaching put pressure on living space. Vast numbers of urban houses – with gas, water and sewers installed – mushroomed in both Britain and the United States between 1880 and 1910. These houses more often than not had insufficient room to accommodate more than one live-in maid. In 1892, the fictional *Diary of a Nobody* by the brothers George and Weedon Grossmith

Triumphs of hope over experience? A mangle (*above*) and a knife polisher (*below*) driven by electric motors.

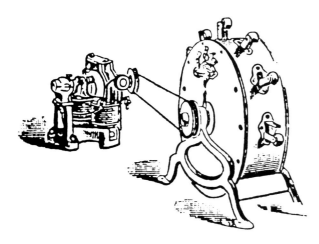

addressed this very phenomenon, suitably endowing their hero Mr Pooter with both bathroom and bath – even though the bath was not of the best quality.

The *Girl's Own Paper* ran a series of articles in 1881 aimed at these very people, entitled: 'Margaret Trent and How She Kept House'. Margaret Trent, as employer of worthy servant Ann, was expected to work alongside – or at least to complement – the said Ann. If she were also to have time to spare to fulfil her social functions, she would need to use labour-saving appliances such as washing machines and gas fires. Husbands too were encouraged to review their situation in life. As Mrs James George Fraser said in *First Aid to the Servantless* in 1913, a husband had to show consideration in a servantless house. 'He need not splash in his bath tub like a hippopotamus at the Zoo', and he should put away his own clothes.

Yet the period from about 1880 until 1930 – when numbers of servants were declining rapidly – saw no corresponding upsurge in domestic inventions nor, more significantly, a notable increase in consumer purchasing – at least in Britain. That had to wait until after the Second World War when all services were commonly available. By then, scientific expertise had caught up with technological needs. Cheap, light metals, easy-to-care-for coatings, electrical switches and electrical timers were all at the service of the mass-producing domestic appliance industry. Wages were in advance of costs. The time was ripe for an unprecedented purchasing bonanza.

If we had to choose one device which had the most far-reaching consequence, it would have to be the small electric motor. Its development from Michael Faraday's first experiments with the 'magneto-electric force' in 1831 was surprisingly slow. In 1889 Nikola Tesla, together with the Westinghouse Company of Pittsburgh PA, finally marketed a 1/6 horse-power motor to drive a three-bladed fan. The time was now ripe for the application of such a motor elsewhere. Not until devices were driven by an electric motor rather than being cranked by hand did household technology begin to be truly labour saving.

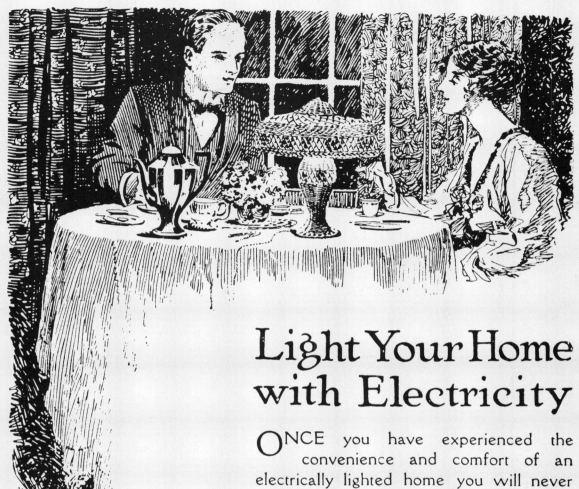

Light Your Home with Electricity

ONCE you have experienced the convenience and comfort of an electrically lighted home you will never go back to any other form of lighting. Electricity transforms the most ordinary house into an exceedingly comfortable and convenient home in which the servant problem is in a great measure solved by a general reduction of housework and a complete elimination of drudgery. Electric Light is the safest, cleanest, healthiest and cheapest of all illuminants.

Issued by the Electric Lamp Manufacturers' Association of Great Britain, Limited.

Electric Light

is a modern necessity

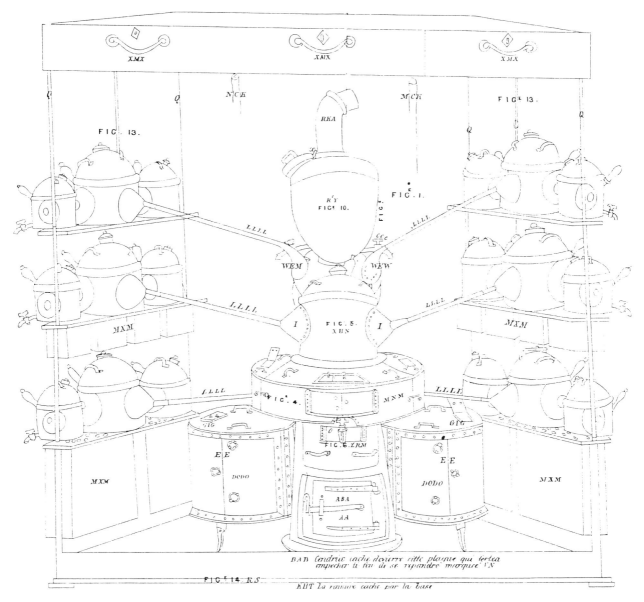

Bartholomew Dominiceti's patent stove (Patent No. 972 of 1770). Dominiceti, a Professor of Physic from Chelsea, describes 'the firestove with boilers, pots and other salutory utensils adapted to the same, which together form a machine called by me The Economist which will be of great utility to His Majesty's subjects in general.'

Cooking

The kitchen range

'To enclose or not to enclose' was the debate which appeared to preoccupy the cognoscenti of the cooking range in the years after the Great Exhibition of 1851.

The open range was basically a refinement of open hearth cooking. A central open fire heated an oven on one side and, more often than not, a boiler on the other. These provided 'hobs' – hot surfaces for pans and kettles – while the open fire was used for roasting – either vertically or horizontally. The spit was turned either by a clockwork jack or by a smoke jack set in motion by the current of air induced by the range. As late as 1925 Gertrude Jekyll wrote: 'Nowadays we roast more conveniently by hanging the joint vertically to the clockwork jack; this is also better suited to the narrow shape of our coal fires.'

Much of the heat went straight up the chimney, and this inefficient use of fuel had been condemned by Benjamin Thomson (better known as Count Rumford, an American member of the British Royal Society) be-

fore the end of the 18th century. However, not until the 1840s were Rumford's principles applied to range design. The size of the grate was reduced and it was covered with an iron plate to direct the heat through flues surrounding oven and boiler. These 'kitcheners', as they were known, offered several advantages: stews could simmer gently over the fire, protected by the hot plate and, for fast boiling, circular holes with removable lids were provided which exposed vessels placed upon them to a higher temperature. Food was no longer in contact with the grime of the fire. Pots and pans were no longer blackened by soot and, being protected from the strongest heat, lasted longer.

It was the question of how best to prepare a good roast that weighed against the closed range. Roasting seemingly needed air; baking did not. To this end, extra roasting ovens were incorporated which allowed hot air to stream through. Dampers, knobs and levers proliferated to allow the maximum use of these in-

An ordinary open-fire kitchen range.

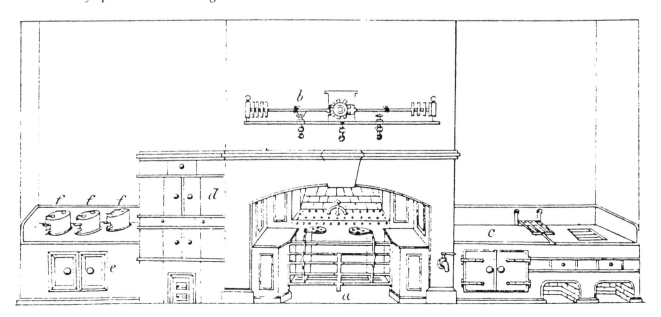

creasingly complicated ranges. Fumes too were a problem. Whereas with the open ranges fumes followed smoke up the chimney, with closed ones they were more likely to accumulate because the chimney opening was adjusted by an adjustable door, or 'register'.

In a badly-constructed – or inexpertly operated – stove, the fumes could be deadly. Picture the harrassed cook wrestling with extremely hot, extremely heavy levers, succumbing to a bout of noxious fumes. Design improvements were made during the latter part of the 19th century but, unless the cook was as proficient a stove manager as she was a producer of food, few economies were made with closed ranges.

An American alternative to these heavy permanent structures was shown at the 1851 Exhibition. The pioneer needed to go West with a cooker: the free-standing closed range. On the grounds of fuel efficiency too, the *Practical Mechanics' Journal* of October 1862 exhorted the wider use and manufacture of 'small cooking stoves and apparatus such as are used in France'. Builders of the British Empire of course were also in the market for portable stoves. Smith & Wellstood were the chief manufacturers and mirrored the need with the evocative names of their models: the *African*, the *Australian*, the *Grand Pacific*, the *Kipling*, the *Lioness*, the *Plantress* etc.

Two innovations promoted the long-term survival of the solid fuel range. The first was the increased use of anthracite where it was available; a hard, slow-

A kitchener – a kitchen range of the mid-19th century.

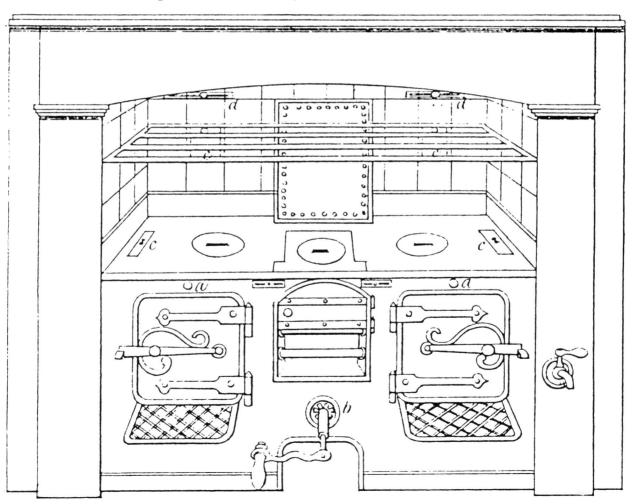

An American cooking stove. The fire is closed. The top forms a hot-plate, and the hot air must pass round every portion of the ovens except the doors before its escape to the chimney.

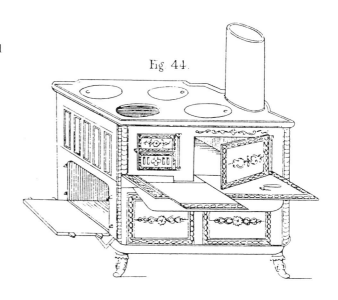

Fig. 44.

The Aga, Gustav Dalen's invention. Introduced in 1924, versions of it are still sold today.

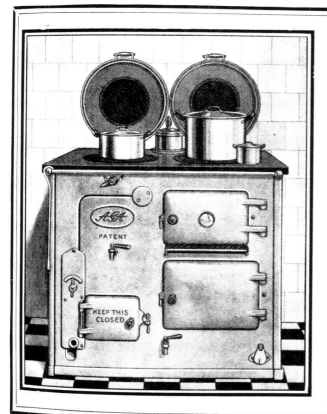

burning coal, anthracite enabled the fire to stay in over night; moreover it produced far less smoke. At the 1882 Smoke Abatement Exhibition, the *Treasure* was shown by its maker Mr Constantine; opening the Exhibition, the Marquis of Lorne was moved to say: 'Were the public but to avail themselves of such appliances, we might one day see roses blooming in Kensington Gardens.' The final seal of approval was given by *Household Management* in 1890; as a result, by the early 20th century, manufacturers were supplying ranges to burn anthracite as well as coal.

The second innovation to promote solid fuel was based on a return to the efficient insulating ideas of Count Rumford. In 1924 Gustav Dalen produced his Aga. The idea was basically very simple: a massive metal fire unit, kept at a very high temperature, which stored, radiated or conducted heat as required. What made the Aga so economical in use was the thick layer of insulating material all around it. It was guaranteed to use no more than one and a half tons of coal a year.

Cooking by gas

The 1851 Exhibition dealt a cruel blow to cooking by gas when the Royal Commission organizing it stipulated that no apparatus could be exhibited which was 'in practical operation through the agency of gas'. Fortunately, those who believed in the future of gas appliances were not deterred.

Gas had been used to illuminate many of the world's great cities; indeed, 215 miles of London's streets had been gas-lit since 1823. Elsewhere in Britain, James Sharp of Northampton gave the first authentic demonstration of cooking by gas about 1830. A certain Mr Russell of Leamington was clearly impressed, declaring publicly that having had a gas cooking apparatus 'in constant use for more than twelve months ... I would not be without it on any account'. Hick's Patent Gas Roaster was produced in 1831, its object being to 'roast meat by the flame of ignited gas, that heat being confined under a conical cover, which is placed as a screen over a circular burner, and the meat to be cooked is mounted on a vertical spit in the centre of a circle of gas flame.'

Sharp's apparatus of 1861 was much more impressive; he claimed that it was able to cook dinner for 100 persons. It was a circular structure with a boiler

Sugg's *Parisian* meat roaster, about 1880.

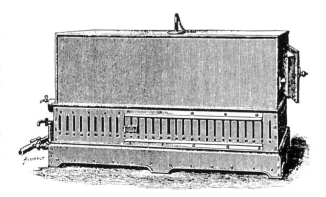

Sugg's *Vienna* bread or pastry oven

'Sharp's gas-cooking apparatus, to be shown in operation at the special exhibition of the Gasfitters' Association, at the Royal Polytechnic Institution', 1851.

standing on top of the oven, on top of which was a chimney open to the room. The whole thing occupied a space four feet six inches (1.4m) in diameter.

The efficiency of combustion was increased by mixing air with the gas. In 1856 Pettit and Smith improved the design by using a row of burners to heat a special material which gave off 'radiant' heat like a fireback – easier to harness than the heat given off by the burners alone.

Gas offered the first really radical technological change in cooking methods. No longer was there fuel to carry, or mess to clear; no rising early to light fires; no constant tending. Yet suspicions about gas cookers prevailed. Many people feared gas would be poisonous. The cookers were also smaller than conventional ranges and the interiors, of tiles or enamel, were not durable enough. The sheet iron of which they were constructed was too good a conductor of heat and it was the end of the 19th century before the problem of insulation was solved.

Many models were on show at the Smoke Abatement Society Exhibition in 1882, but it was William Sugg, a persuasive manufacturer and visionary, who penetrated the prejudice against gas and promoted its use to the next stage. Sugg published a book – *The Domestic Uses of Coal Gas* – in 1884; his daughter Marie Jenny produced a manual of gas cookery in 1890. Sugg felt it necessary to remind his readers that it was unnecessary to have red-hot glowing coals by which to roast, and promised that 'with the aid of gas an ordinary plain cook can roast a joint to a turn.'

Different designs met different needs. There was the *Parisian* roaster and the *Vienna* bread or pastry oven while, for the few ordinary gas-using folk, the *Charing Cross Kitchener* was the epitome of efficiency and utility, combining all the opportunities offered by Sugg's other more specialized apparatuses. In Britain, the gas companies were by now hiring out cookers to interested customers, and once the slot meter system was instituted, increasing numbers of people felt disposed to experiment. The influential *Household Management* now gave its cautiously-enthusiastic blessing to the results of gas cooking. In the United States, the American Gas Association worked hard to improve their cookers – and their image.

As the end of the century approached, William Parkinson and Company of Birmingham, gas engineers

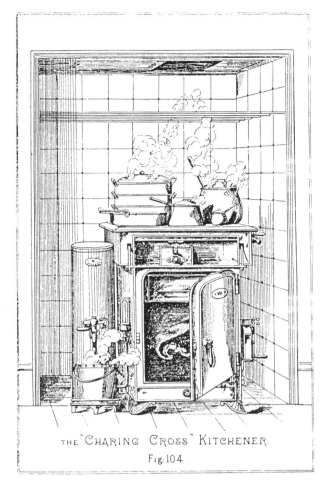

THE "CHARING CROSS" KITCHENER.
Fig. 104.

The *Charing Cross Kitchener* which, Sugg was able to assure users, met the requirements of the gas companies and was 'sufficient to do all the roasting, baking and boiling for any number of persons from six to twenty.'

Right
'The deluxe *Mainamel* gas cooker can be wiped down in a moment and restored to spotlessness without the slightest effort.' Note the Regulo oven thermostat in the pipe on the right.

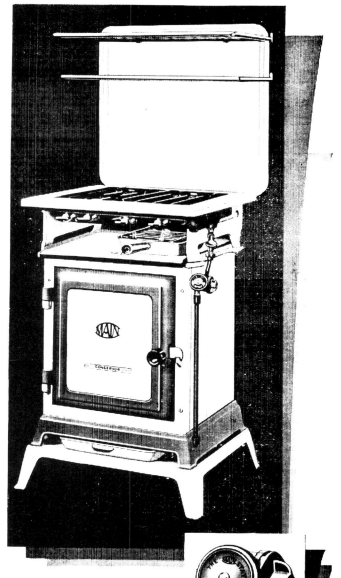

THIS FULLY ENAMELLED GAS COOKER

. . . . . Can be wiped down in a moment and restored to spotlessness without the slightest effort.

The enamelling is smooth as glass and will not crack. Inside and outside, back and front are finished in enamel—the effect is beautiful.

Every modern improvement is incorporated in the design; gas is saved by the patent 'Conservor' oven which bakes on every shelf, and the front reading RED DIAL 'Mainstat' automatically takes control of the food being cooked. The DELUXE MAINAMEL Cooker is obtainable from Gas Undertakings everywhere. Ask for terms at your local Gas Showrooms or write to sole manufacturers for booklet :— "Colour in the Kitchen" and also free Recipe Book.

R. & A. MAIN LTD.

Department 40, GOTHIC WORKS, EDMONTON, LONDON, N.18 and GOTHIC WORKS, FALKIRK.

London Showrooms: 48 GROSVENOR GARDENS, S.W.1 *Glasgow Showrooms :* 82 GORDON STREET.

Sole Manufacturers of the CINDERELLA Inset Gas Fire

LOOK FOR THE NAME **MAIN** ON THE DOOR PANEL

DELUXE MAINAMEL *Gas Cookers*

THE MAINSTAT Automatic Cooking Control with the RED Dial ensures correct temperatures being maintained and repeated.

Margaret Fairclough cooking with electricity at her Gloucester Road School of Cookery, London, 1895.

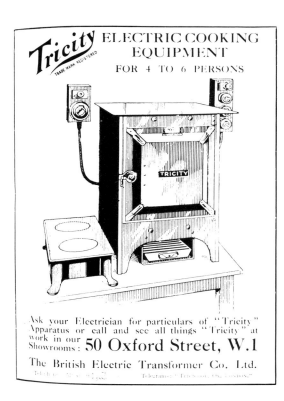

An early post-war Tricity cooker. Note the switches on the wall for the different cooking functions.

since 1816, moved into the field of cookers. Models became properly ventilated, and had removable enamelled linings and fittings to make cleaning easier. Grills too began to appear; by 1914 gas cookers were well accepted and had assumed designs more akin to the modern cooker than to their contemporary black ranges.

Lighter enamelled sheet steel became available after the First World War; moreover, houses both old and new were now being piped not just for lighting, but for cooking. The appearance of the Regulo oven thermostat in 1923 led to an increasing popularity for gas cookers. By 1930, they prevailed all over the United States and, by 1939, nine million were in use in Britain.

Cooking by electricity

As with gas, electricity was at first used only for lighting, and it was some time before electric cooking appliances were designed. Attempts were made to interest the cooking public at the Electrical Fair at the Crystal Palace in 1891. The Columbian Exhibition Centre in Chicago included a Model Electric Kitchen in 1893 and in 1895 an electrically-cooked banquet was given in honour of the Lord Mayor of London.

Crompton and Company retailed the first electric cookers in 1894 in Britain. The elements were of steel wire coiled around ceramic cylinders. However, the steel was liable to rust and crack at too great a heat – hardly a recommendation in a cooker. In 1906, L March of Illinois began using wire of an alloy of nickel and chrome – which he called Nichrome – which gave out more heat than steel, and was far more durable. By 1912, C R Belling had developed a ceramic material to support a coil of nichrome wire; when heated to incandescence, it gave off considerable heat.

Originally the cooking elements were placed on either side of the oven, but they were eventually found to be more efficient if placed top and bottom. The *Beeton* of 1906 still had its elements at the side. It had no hot plate, but provided instead a socket for plugging in

Right
'The value of good and varied cooking is becoming more appreciated by the housewife of today. At Olympia [The Ideal Home Exhibition, 1935] she will see many and varied types of cooker which can make cookery child's play'.

The IMPORTANCE *of* GOOD COOKING

Cookers of varied types.

THIS STEEL-BUILT semi-insulated cooker will cater for up to 15-18 persons. Practically the entire hot-plate can be used for boiling purposes. Crittall Cookers, Ltd.

The "Super Interoven" stove is on view at the stand of the Interoven Stove Co. It cooks for as many as nine persons, baking joints, cakes, bread and pastry to perfection. It also gives an ample supply

THE value of good and varied cooking is becoming more appreciated by the housewife of to-day. At Olympia she will see many and varied types of cookers which can make cookery child's-play. Every feature of successful cooking by gas, electricity, oil, and coal is shown.

An interesting double oven, "Butto," is shown by Crittall Cookers, Ltd. This cooker is a steel-built, semi-insulated model designed on scientific lines for high efficiency and low fuel consumption. It will cater for from 20 to 25 persons, according to the type of menu desired. Another model exhibited is illustrated on this page.

THE LATEST model of the "Foresight" patent grates for cooking and hot water supply is illustrated above. The upper doors enclose a hob which goes the whole width of the grate.

DESIGNED to ensure a high degree of flexibility, the "Eagle" coke range (shown above), is well suited for large and medium-sized households.

PROVIDING separate oven and boiler flues, the new "Tweenie" grate is ideal for the small kitchen where space is a consideration. It is thoroughly efficient in every working detail and simple and clean in use. Triplex.

THE new model of the Super "Kooksjoie" range is heavily lagged, which increases the efficiency to such an extent that the fire-box has been greatly reduced in size with a resulting reduction in fuel consumption. London Warming Co.

"SENTRY" combination boiler and oven is made in various sizes to suit any house. Constant hot water and efficient cooking is obtained from one fire. The white enamel side panels and tiled surround are very easily cleaned.

of hot water at only half the amount of fuel consumed by an ordinary kitchen range.

The London Warming Co.'s "Super Kooksjoie" range can be seen on the stand of the Amalgamated Anthracite Collieries. The "Kooksjoie" is an excellent kitchen range, and this new model has the advantage of being heavily lagged, which increases the efficiency to such an extent that the firebox has been greatly reduced in size, with a resulting reduction in fuel consumption.

Thermostatic oven-heat control is provided; all cooking operations, from quick boiling to simmering, can be carried out on the hotplate; and a continuous supply

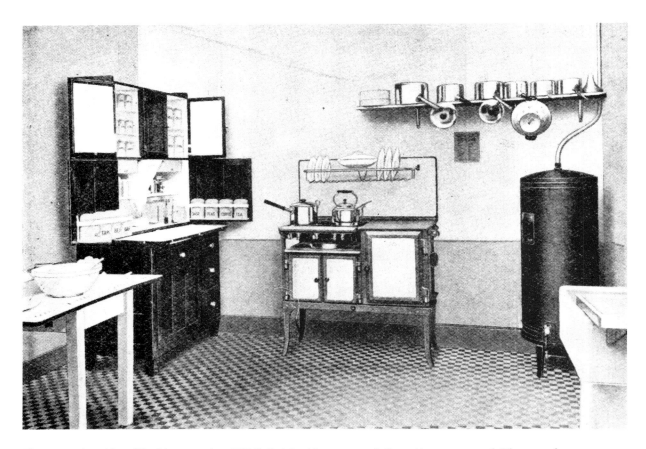

The up-to-date New World gas cooker (1930) finished in wear-resisting vitreous enamel. The oven has a low-temperature zone so that high and low temperature cooking can be done at the same time without attention. Note also the 'Sunhot' gas boiler, 'which can be fitted without any structural alteration'.

an electric kettle or frying pan. *Household Management* in 1907 suggested, with justifiable caution, that electricity was 'quite practical'.

Few appliances were bought – but then, few homes were wired for electricity. Appliances were made of cast iron, so were neither easier to clean nor more attractive than those heated by gas or solid fuel. Nor, in fact, were they particularly efficient, the chief complaint being their lack of speed. It could take up to 20 minutes to boil two pints of water and 35 minutes to pre-heat an oven. Moreover, electricity was still distrusted as a fuel which could be neither seen nor smelled.

Family-sized cookers were still too expensive for most people. Mindful of competition, British electricity companies adopted the same procedure as the gas companies had done and hired out their ap-

pliances. This paid off; in Manchester, for example, sales began to outweigh loans in 1929 as generation and distribution technology developed and new houses were wired for electricty. The Creda cooker, which allowed the heat to be properly controlled, was launched in 1931. It was the first to embody a thermostat and, by 1939, at least one million British customers for electric cookers agreed on their cleanliness and efficiency. The figures were considerably higher for the United States, where prices had been falling throughout the 1930s.

Right
The latest in gas cookers, 1939.

The popular choice for the kitchens of to-day

Housewives everywhere are delighted with the smooth, white enamelled cabinet-like appearance of the Parkinson "RENOWN" Gas Cooker—it has indeed proved to be the cooker of their dreams. Thousands, too, are finding that balanced heat cooking, a result of the modern oven design of the "Renown" Gas Cooker, appeals as much as the attractive appearance of this cooker. At a modest cost, you, too, can have a "RENOWN."

Visit your LOCAL GAS SHOW-ROOMS and ask for a demonstration.

SEE THIS WONDERFUL COOKER ON STAND No. 137 Ground Floor, IDEAL HOME EXHIBITION, EARLS COURT.
April 11—May 6

POST THIS COUPON NOW
For Parkinson "Renown" Brochure and FREE Recipe Booklet.

Name ....

Address ....

....

....

I.H.4.39.

The PARKINSON
RENOWN
GAS COOKER

THE PARKINSON STOVE CO., LTD., BIRMINGHAM, 9. London Showrooms : Terminal House, Grosvenor Gardens, S.W.1

The bathroom

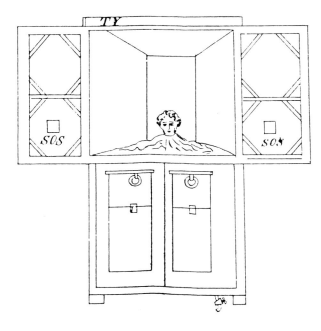

Dominiceti's versatile stove (see page 10) was provided with a Turkish bath module.

Baths and showers

The key requirement for a leisurely soak in the long bath that we know today was a sufficient supply of hot water; second to that was a means for its subsequent disposal. The road to modern bathing habits was not straightforward. In some quarters – even late in the 19th century – hot baths were viewed with suspicion and recommended chiefly for invalids.

Whether the lack of enthusiasm for hot water was a result of its shortage or of strongly-held views is not clear. In the 1870s, *Cassell's Household Guide* described the cold bath – one at a temperature around $60^{\circ}F$ – as 'one of the most refreshing comforts and luxuries of life ... calculated to make the whole body rejoice with buoyancy and exhilaration of spirits.' That steam or vapour baths were also popular is demonstrated by the rise in the number of Turkish-style public baths during the latter half of the century.

Domestic versions, such as those offered in the 'Health Department' of the 1902 Sears Roebuck Catalogue, or in the 1907 Army & Navy Stores Catalogue, were available and, more to the point, needed only small amounts of water – making them more akin to saunas than steam baths.

Like steam baths, showers needed only a small water supply. The earliest patent for a shower (1767) was taken out by William Feetham, and a wide range was available by the time of the Great Exhibition. Early showers were usually hand pumped and could be fitted over any shape of bath tub. The water in the reservoir was released at the pull of a handle and water poured through perforations in its base.

In 1879, Warren Wasson and Charles Harris of Carson City, NV, patented a shower which required the bather to keep a constant treading movement to operate a pump which circulated the same supply of water. However warm it was at the outset, it doubtless soon became cold, and therefore – making a virtue of necessity – invigorating. It would have been more

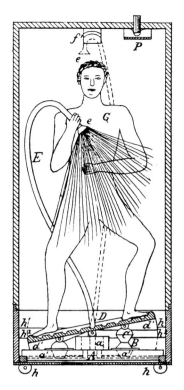

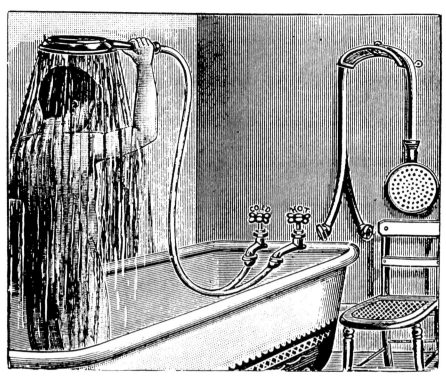

Left

Wasson & Harris's patent pedal shower (US Patent No. 327 of 1877).

Right

'Bath shower consisting of Indiarubber Tube with Nozzles for "Hot" and "Cold" Taps and Nickel-plated Hand-shower. . . . The hot and cold water can be mixed to any temperature . . . they will not spray the walls of the Bath-room . . . They give no disagreeable shock, but are pleasant and agreeable, so much so that the most enfeebled can use them with comfort and safety.' 1903.

inviting in Nevada than in colder regions. William Luther's patent (1891) illustrates a portable shower in which a pump compresses the air in a reservoir to force the water up to the shower head. Other designs, wherein the bather wore a necklace of pierced rubber which could be attached to a tap, were more certainly cold – 'a great life invigorator' commented Sears Roebuck in 1902.

The ultimate in shower designs was incorporated into the Hooded Bath. The user could inject water from all angles by manipulating a range of knobs. Later inventors combined 'plunge, spray, shower, douche and wave'. Ewart's Improved Spray Bath of 1882 has no fewer than 10 controls, though simpler versions were offered from 1900.

Showers declined in importance from the 1870s as piped water and waste disposal became common in middle-class houses, paving the way for the bathroom as we know it. Most people were used to the tub in front of the fire and no doubt it was the inconvenience of heating, filling and then disposing of so much water that made bathing such an infrequent occurrence.

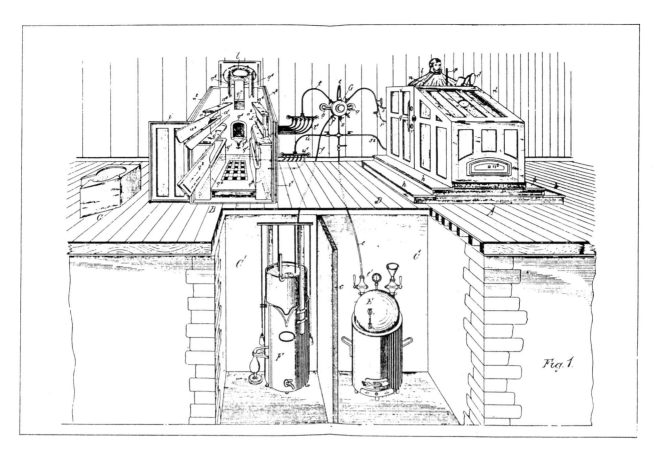

Charles Berthe's patent safe hot air or vapour bath (17477 of 1890) was designed to eradicate the 'numerous fatal accidents', usually by 'fire or asphyxiation'. Berthe was manager of the Jersey Hydropathic Establishment and Medical Gymnasium.

Heating the water

One way around the hot water problem was to fill the bath with cold water and heat the bath itself. Defries in the *Journal of Gas Lighting* claimed that his Magic Bath Heater would heat 45 gallons of water to 106°F 'in the incredibly short period of six minutes – the cost of the gas being little more than one penny.' This heater – a burner set under a metal bath – was an 'indispensable requisite for the comfort and preservation of life. Its simplicity [placed it] within the management of a child.' In 1851, William Tyler exhibited a bath which was similarly heated by coal. In the same year, the Pre-Raphelite artist John E Millais was in his London studio, painting Lizzie Siddall as Ophelia lying in a bath of water 'kept hot by lamps underneath'. It must have been like sitting in a kettle on a stove.

What a boon it would be if a supply of instant hot water were constantly available! In 1868, Benjamin Waddy Maughan patented the gas water heater to which he gave the name 'geyser' after the Icelandic hot springs. It was followed by Sugg's Boiling Stream Therma. In these early geysers, the water merely trickled or sprayed down among the flames from gas burners. In later designs, for example The New Rapid from Fletcher Russell Co, the water ran through a pipe which took a spiralling course through the hot gas from the burners. Not only did this heat the water more efficiently; it also saved it from contamination by exposure to the gas fumes.

Early geysers generally lacked safety devices and it was important to follow complicated instructions to prevent accidents, but in time they became an extremely popular form of water heating for both basins and baths. At the 1882 Crystal Palace Exhibition, Strode showed a design which, he claimed, could heat 80 gallons to 60°F using gas at the rate of 100 cubic feet per hour. These geysers were designed to operate at the points of use. But in 1899, Ewart produced the Califont, which could 'be fixed in any out of the way part of the building where most convenient', and produced enough hot water – at temperatures up to 200°F – to feed every tap on the circuit.

By the late 1860s it was becoming the practice to build hot water boilers into kitchen ranges. In 1869, Comyn Ching patented a boiler shaped to form the back of the grate. However, a spate of explosions in kitchens was reported in December of that year, the result of the hot water cylinder being placed near the top of the house with no safety valve and the companion cold water tank freezing. Despite a subsequent plethora of patents aimed at remedying the problem, not until the 1930s was the back boiler a common and reliable way of providing hot water for the home.

The availability of hot water upstairs led to the development of the bathroom proper. In the last 20 years of the 19th century, plumbers and furnishers consorted to make the bathroom – in wealthy homes at least – a central feature, rather than the attachment to the bedroom it had been previously.

Bathtubs were of two classes – copiously-decorated porcelain for the well-to-do, and cast iron for the aspiring. Cast iron could be galvanized, painted or enamelled, though none of these finishes was long lasting. Until 'porcelain enamel' – a hard coating which could expand and contract along with the bath – was perfected, many early do-it-yourselfers must have run

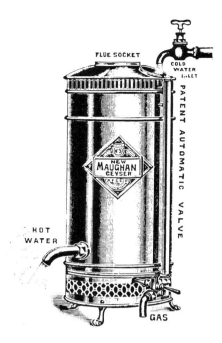

Maughan's Geyser, 1901 pattern.

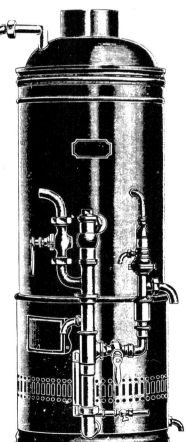

The Califont.

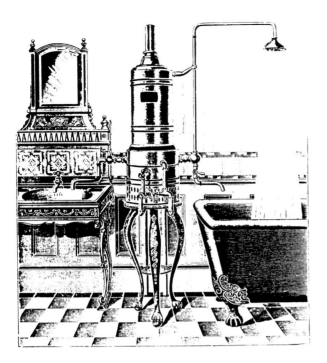

The Acme geyser.

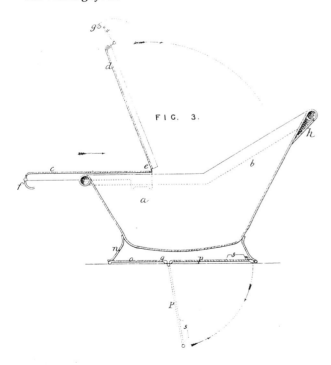

Vallas's patent bath for use as a travelling trunk (3152 of 1861). Articles may be stored both in the main bath (whose top slides off as shown) and in the base.

into the same problem as the fictional Mr Pooter when he painted his bath with red enamel – only to find it dissolving in the boiling water a couple of days later.

Bathrooms appeared in large American homes – and even hotels – decades earlier than in their European counterparts. Although totally functional in style, in the 1850s the Mount Vernon Hotel at Cape May, New Jersey, predated the Paris Ritz by some 50 years in boasting a bath with hot and cold running water for every bedroom – at least, according to the *Illustrated London News* reporter.

The recognition for space saving in bathrooms percolated only slowly into private houses in Europe. Relieved, presumably, of the continuous urge to travel westwards, Europeans did not embrace the American penchant for folding tubs as offered in the Montgomery Ward mail order catalogue of 1895.

The closet

In retrospect, one of the more interesting features of the Great Exhibition was its provision of public water closets – toilet facilities with a water flush and waste system, albeit rudimentary. In fact they were used by only about 14 per cent of visitors; what the other 86 per cent did is a matter for conjecture. The general public retained a curious antipathy towards public lavatories – where they were provided, nearby residents often petitioned for their removal.

The need for 'portable water closets or latrines for fêtes, racecourses, exhibitions or other places where the public temporarily congregate for amusement' was met by André Cassard of Paris. There was a problem with such structures then, just as there is today, a fact recognized by the Glasgow ironfounder John Millar, who patented a 'method of rendering urinal plates incapable of being written upon'.

If people avoided visiting the WC, it may have been because early models were unequivocally foul smelling. Most people would have used a wooden seat with a hole in it directly over a cess-pit.

In 1860, the Reverend Henry Moule patented an earth closet: a pull handle released a measured quantity of 'dry peat, earth, lime, clay &c' on to the contents of the bucket. In spite of a host of other earth closet patents, Moule's system continued to sell until the Second World War. 'The charge of earth is delivered by

George Vanderbilt's bathroom, 1885 – one of the first in a private home.

Mount Vernon Hotel, Cape May, New Jersey. It offered a bathroom with every bedroom.

PLUMBERS,

And Manufacturers of Barrows' Patent Cooking Range,

pulling the handle forward', or up, or – in the de luxe model – 'automatically, upon the user rising from the seat'. The contents of the bucket would be removed at intervals by night-soil men – midden men in some areas.

With the same principle, but different droppings, the Triumph Peat Dust closet appeared in the 1890s and claimed superiority over earth and ashes – 'Cheaper and Nicer. No dirt, no smell' – or so said the

A fanciful plumber's advertisement from the Boston Directory 1850–51.

Right
A range of Moule models from Winstone's builders' catalogue about 1905.

No. 12 MOULE'S PATENT EARTH CLOSET.

32/-

COMPLETE WITH GALVANIZED RIM.

Pull-out Apparatus, without Seat or Bearers.

The charge of earth is delivered by pulling the handle forward.

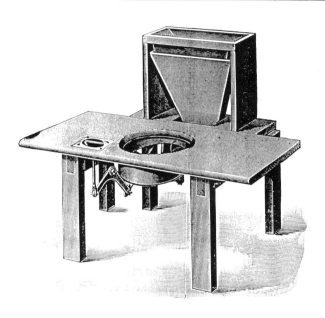

No. 13 MOULE'S PATENT EARTH CLOSET.

48/-

COMPLETE WITH GALVANIZED RIM,

SEAT AND BEARERS

The charge of earth is delivered by pulling up the
handle in seat.

The Seat is 3 ft. long, and can be cut to the length required.

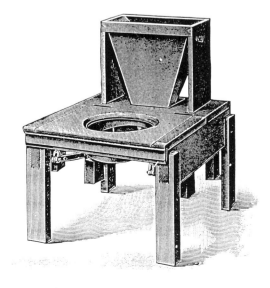

No. 14 MOULE'S PATENT EARTH CLOSET.

60/-

COMPLETE WITH GALVANIZED RIM,

SEAT AND BEARERS.

Self-acting apparatus. The charge of earth is delivered
automatically, upon the user rising from seat.

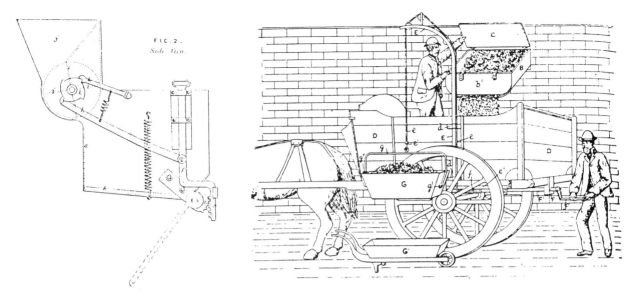

Revd Henry Moule's patent earth closet (Patent No. 1316 of 1860). The drawing shows delivery mechanism: when the handle is pulled, the pawl rotates the drum a quarter of a turn, delivering a measured quantity of material.

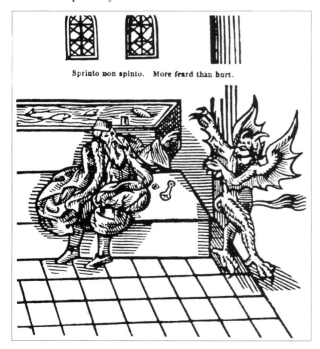

Sprinto non spinto.   More feard than hurt.

Sir John Harington *The Metamorphosis of Ajax* : A devil speaking to 'a godly father sitting on a draught'. The first depiction of a water closet.

advertisements. Dry systems such as these were recommended as more convenient in hospitals, bedrooms and nurseries well into the 20th century, while the combination of corporate house-building and a lack of suitable sewerage facilities meant that for a long time there was no alternative to dry systems in working-class homes.

An improved means of disposing of human waste was highly desirable, and flushing with fresh water seemed the most obvious approach – at least to sundry inventors, if not to planners of public water systems.

The first recorded water closet was that of Sir John Harington in 1596. Although various English designs are to be found in 18th-century France (les affaires anglaises), no patents were taken out until that of Alexander Cumming, a watchmaker of Bond Street, in 1775. Cumming's design, modified only slightly by Joseph Bramah in 1778, was to hold sway for almost a century.

These early closets had two valves: one to admit water to flush, the other in the base of the pan to control the outlet. This lower valve was, after Bramah's modification, hinged; the idea was that some water should be retained in the pan after flushing. The action was fairly vigorous but not always sufficiently forceful to dispose of the contents, and the lower valve did not always provide an efficient water seal. Consequently, a little-used closet, where the remaining water could evaporate, or a closet with an especially ill-fitting valve, would give rise to fairly anti-social smells.

Night-soil men visiting John Wood's patent midden (Patent No. 5518 of 1887).

The curious thing is that Cumming's patent clearly shows a water trap – an S-bend down the soil pipe. If this had been moved up to replace the flap valve, the modern pan would have been born a century earlier.

Close on the heels of the water closet came the flushing cistern – or water-waste preventer – to deliver a calculated quantity of water. 1870 was, as Lawrence Wright puts it in his book *Clean and Decent*, the *annus mirabilis* – or remarkable year – of the water closet. T W Twyford of Hanley produced an earthenware wash-out closet: water washed around the basin and some was left in the basin to help the flushing water remove the waste. In the same year, Hellyer's Optimus, an 'improved valve closet', boasted a more reliable handle and a more efficient water seal to help keep the smells down. Hellyer's closets had casings of mahogany or wicker to conceal the workings; the effect was to make the structure increasingly throne-like.

Also in 1870 J R Mann produced his siphonic closet. After the initial fast flush there followed a slower

The first patent for a water closet: Alexander Cumming (Patent No. 1105 of 1775).

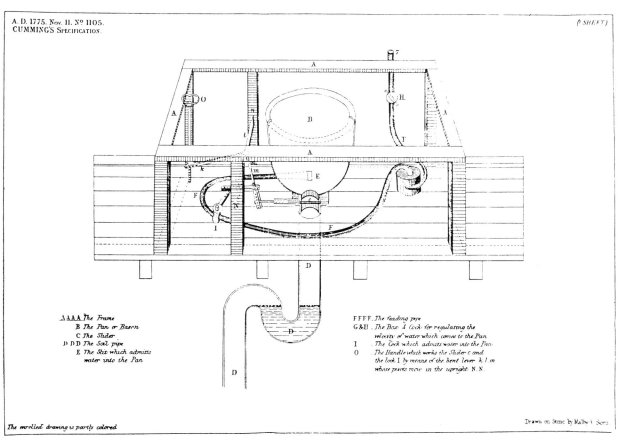

A. D. 1775. Nov. 11. Nº 1105.
CUMMING'S SPECIFICATION.

(1 SHEET)

A A A A The Frame
B The Pan or Bason
C The Slider
D D D The Soil pipe
E The Slit which admits water into the Pan

F F F F The feading pipe
G & H . The Box & Cock for regulating the velocity of water which comes to the Pan
I . The Cock which admits water into the Pan
O . The Handle which works the Slider c and the lock I by means of the bent lever k l m whose parts move in the upright N N

The enrolled drawing is partly colored

Drawn on Stone by Malby & Sons

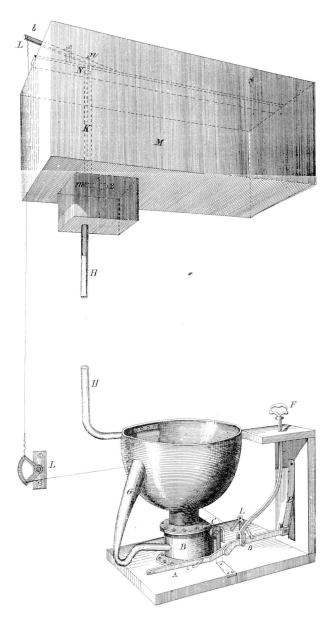

The second patent for a water closet: Joseph Bramah (Patent No. 1177 of 1778). Pulling handle F opens the flap valve at B and actuates the cord which releases water from a header tank.

secondary flush to keep things moving. The complexities of the system prevented it becoming as successful as the washdown type, despite the fact that it was considerably quieter.

At the 1884 Health Exhibition Jennings' Pedestal Vase won a Gold Medal. Although name and function were incongruously allied, the judges were clearly impressed when, with a two-gallon flush, it cleared:

Ten apples averaging one-and-a-quarter inches diameter
One flat sponge about four-and-a-half inches diameter
Plumbers' smudge coated all over the pan
Four pieces of paper adhering closely to the soiled surface.

Finally, in 1887, D T Bostel of Brighton patented the washdown water closet in its present form – incorporating an S-bend. This did away with the need for a risky valve, as the water retained in the S-bend became the trap. From then on, interest moved away from technology to decoration; the years up to the First World War produced a feast of creativity in sanitary design.

This did not, of course, deter the intrepid from patenting what they saw as refinements – brought on as usual by some personal experience; A D Lagrelle, for example, patented an electrically-heated seat (1893). However, the current was switched on when the user sat down, so the seat was probably well warmed before the electrical heating took effect.

The last word was perhaps R C Sayer's improvement which was designed to burn any solid matter deposited in the pan to a fine ash. This process necessitated a supply of 'compressed air, combustible gas, and exhaust and drain pipes laid on to each closet in the district' and could hardly, if widely adopted, have contributed a great deal to safety in the home.

Between the wars, piped water and sewage provision continued to extend into the more densely-populated city areas and sparser rural ones, thus enabling more households to benefit from fitted baths and water closets.

The home laundry

Washing machines

Short of having someone else to do the work, those whose business it is to see to the laundering need a machine which can fill itself with water at the appropriate temperature, rub the clothes clean, drain itself, wring the clothes, tumble dry and iron them – and then put them away. A vision perhaps?

Early attempts at relieving at least some of the drudgery of washing clothes were aimed at large laundries and the textile industry, which could use an existing water or steam power source. This was reflected in the exhibits at the 1851 Exhibition where the laundering machines were, in the main, commercial. For example, Mr MacAlpine's machine, developed 'after 20 years of experience and eight months of experiments ... is driven by steam but the patentees intend to construct them of a smaller size to be worked by hand.'

In a small establishment, the laundress used a 'dolly' or 'posser' (from the Middle English *poss*, to pound or beat) to agitate the clothes, a washboard (in use from about 1860), or a Canadian washer – a copper cone with holes to replace the cluster of legs of the dolly. The design of the pierced cone was such that the water was fiercely agitated, helping to scour the dirt from the laundry.

✳ Some 2000 patents for machines or aids to washing were filed in both Britain and the United States in the 20 years following the 1851 Exhibition. Some machines had a dolly attached to the lid which churned or rotated the wash; some had a gyrator attached to the base of the tub; in some the drum itself rotated, causing the wash to be rubbed against the sides or the bottom or even against rollers. Whatever the action, few had any merits. They had in any case to be filled by hand with previously-heated water and then cranked by hand, the contents tending to wind around the legs of the dolly or knot together.

There were many attempts to solve these problems. S S Shipley's model of 1855 used gas burners to heat water in an outer tub, while the wash was contained and rotated in an inner, perforated one. Another model, of 1860, used a back-and-forth motion to minimize the tangling. At about that time, Chatterton

Mr MacAlpine's washing machine.

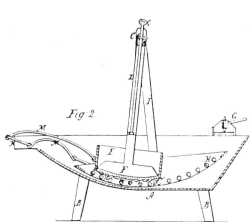

Two early US patents for washing machines – B Hinckley (1831) and E D Wilson (US Patent No. 4891 of 1846).

& Bennett's Float Washer was produced in Manchester, imitating 'exactly the knuckles of a vigorous washerwoman' according to the *Journal of Domestic Appliances*. The design had some merit for it was still to be found in 1926 in the American Montgomery Ward mail-order catalogue.

Less abrasive in motion were the Vowel machines developed by Mr Bradford of Saltram – indeed, in 1883, *Household Management* praised his 'E-type' design. The journal agreed: 'There is nothing in the Vowel which can destroy clothes in any way'. It was of course a moot point which could harm clothes most – an inefficient machine, a vigorous washerwoman, or the boiling water itself. Interestingly, the only washing machine to be featured in the Army & Navy Stores' Catalogue of 1907 is the Vowel 'A' costing 63s9d (about £3.20 or $5). Similarly kind to the wash was Faithfull's

Cradle which worked – as the name suggests – on a rocking principle, producing a figure-of-eight movement in the wash.

Shipley's design of 1855, which mirrored the 1851 design by the American J T King, was a forerunner of some modern washing machines and lent itself most readily to having the water heated inside the tub. J T King was adamant that there was no need for 'rubbing, pounders or dashers' and that the mere presence of steam would neutralize 'the oily, glutinous particles by which means dirt adheres to the fabric'.

Pearson's Marvellous Steam Washer of 1862 was heated by charcoal or gas; Morton's model of 1884 had a built-in gas burner, and as late as 1923 the German firm of Krauss produced a model with a coal fire fitted under the boiler.

Many still had to be convinced of the merits of a

Hibbard, Spencer, Bartlett & Co's catalogue, Chicago 1899. Section of the Wayne washing machine; 'Family size $48 per dozen, Large size $58 per dozen'!

Bradford's Vowel design, about 1880.

The Household Washing Machine: W Summerscales
& Sons Ltd, Keighley, Yorkshire, 1905. The gearing is
arranged so that the handle will drive either the
agitator or the wringer.

J T King's machine (US Patent No. 8446 of
21 October 1851).

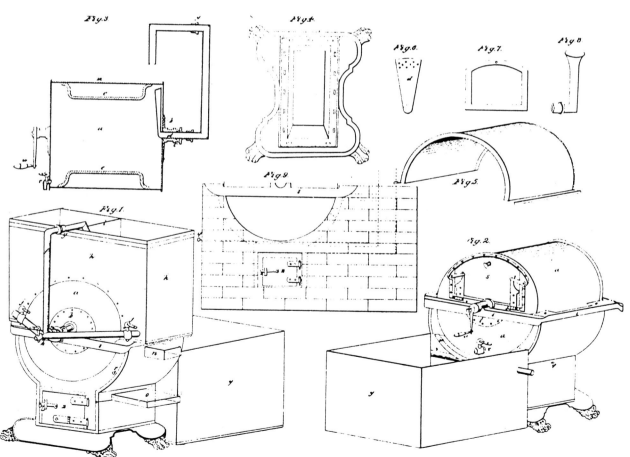

machine for laundering, as advertising copy in the 1902 Sears Roebuck catalogue reveals. 'Absolutely no danger of rust spots on clothes' claimed one, while another was 'warranted not to leak'. Apart from the fact that the laundress could be relieved of the task of manually pounding in boiling soapy water, these early machines had few other advantages. They still had to be filled and emptied and, in between, turned by hand. The same catalogue reminded its readers:

Soak your clothes the evening before washday and soap dirty parts well. Put the necessary amount of clothes to be washed and add a wash boiler full of hot soapy water or enough to cover the clothes thoroughly. Operate the machine about 10 minutes. Take off the dirty water and fill the machine with clear water and operate the machine about 2 or 3 minutes and the clothes will be rinsed.

It is hardly surprising that, until the end of the 19th century, the domestic washing machine was of marginal importance in America because of the growth of public laundries. It was the electric motor that finally made the washing machine domestically successful. Early electrical attachments were crude, being seldom earthed – grounded – and often attached to the underside of a machine where they were prey to dripping water from a leaky tub.

Contemporary with Fisher's machine (*left*) was another with an integral motor, the Thor from Chicago. This also had a powered wringer; it was made of enamelled wood and could be plugged into an electric light socket. But motors were still being mounted beneath the tub, as in the Red Star by Beatty Brothers of Canada (1914). Other machines were beginning to be made of cast iron or steel.

The impetus the First World War gave to the domestic appliance industry in terms of new, lighter metals and longer production runs, enabled the American market for washing machines to boom in the 1920s. In fact the status of the washing machine probably epitomized the gulf then existing between the 'average' British and American household. The United States had solved its electricity supply problems before the outbreak of the First World War. With piped water becoming increasingly common through the 1920s, it is not surprising that 84 per cent of all new washing machines made in 1929 were electric. Paradoxically, more washing was sent out to public laundries in the United States. The practice had peaked before the

A J Fisher's 1909 American machine was one of the first incorporating an electric motor. At last, the washerwoman's arms were no longer needed to agitate the drum. (Patent No. 22114 of 1909).

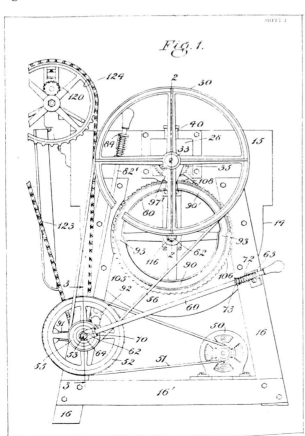

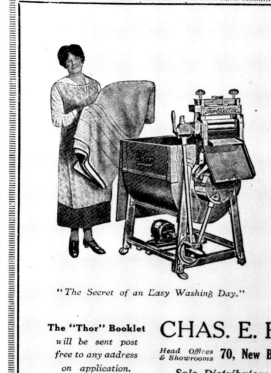

The secret of an easy wash day – plug the Thor into the electric light and the machine will leap into action!

onset of the Depression and finally gave way to the efficient competition from the domestic washing machine.

A patent filed in 1920 in Britain by the American Laundry Machinery Company offered an appliance with the means of automatically controlling the supply of hot and cold water, the drain valve and the supply of detergents in 'definite quantities according to a predetermined cycle. Alarm bells and lights signify when a set of operations is complete.' The 1927 Thor version revealed a multifunctional approach par excellence. The Electric Servant incorporated an iron, food mixer, stirrer, masher – and a wireless.

British developers were making advances and could produce 'automatically arranged "rest" periods' but demand was slower and later. The really basic problem of providing ready supplies of piped hot water had not been solved by the time the Second World War broke out. Indeed, the Heating of Dwellings Inquiry of

AN ELECTRIC HOME IS HEALTHY, CLEAN AND COMFORTABLE

USE

Hotpoint

REGD. TRADE MARK

ELECTRIC APPLIANCES
in YOUR home

MADE IN ENGLAND

MODEL 6012 T.S.
Suitable for 4-6 persons.

The Electric Cooker

makes no fumes or dirt.

It is easy and economical to use and gives the best results obtainable.

The wonderful Torribar Radiant Boiling Plate glows red, heats rapidly and does not require special utensils.

The Electric Cleaner

prolongs the lives of carpets, and rugs.

It is fitted with a power-driven soft bristle brush, and will remove deeply embedded dirt and grit as well as surface dust.

List price only **£10-17-6** which INCLUDES a complete set of dusting tools.

The Electric Washer

means the end of drudgery.

It will complete the weekly wash for, say, a family of four persons in an hour.

It will wash clean the dirtiest garments without fear of damage to the daintiest lingerie.

For particulars of other Electric Appliances such as Irons, Kettles, Toasters, Fires, etc., see your local Supply Authority or Dealer or apply to :—

The Hotpoint Electric Appliance Co. Ltd.

Head Office: **24 NEWMAN STREET, LONDON, W.1**

Telephone — MUSEUM 9144/9.

HOTPOINT HOUSE, 99, CORPORATION ST., MANCHESTER

Telephone—BLACKFRIARS 8441/3.

Also Branches at **Birmingham, Bradford, Bristol, Dublin, Edinburgh, Glasgow, Newcastle-upon-Tyne, Norwich, Nottingham, Southampton.**

Left
A Hotpoint electric washer meant 'the end of drudgery'.

1942 found that 75% of households still had to heat their washing water specially.

The national electricity distribution network in Britain spread but slowly, which did little to encourage an enterprising interest in electrical appliances for the public generally. A Hotpoint was chosen in 1935 for King George V's Jubilee House. In addition to its suitably deferential action – 'cleansing the dirtiest garments yet without fear of damage to the daintiest of lingerie' – it had adjustable legs to suit the height of the owner. This sort of publicity, which implicitly emphasized the gap between the 'could haves' and the 'could not haves', must have served only to make more remote the appreciation that a washing machine could be an effective servant in the homes of ordinary folk.

In Britain, this appreciation did not generally dawn until the affluent decades after the Second World War,

A box mangle.
(*Science Museum*)

by which time sophisticated automatic control devices freed the laundress to go and do something else.

Mangles, wringers and irons

Mangling dates from the beginning of the 18th century, and then referred to the process of smoothing dry or damp linen. Later, the word came to describe forcing water out of wet linen. Mangles were of two designs. The most common was the box mangle. The box – some six feet long and four feet wide – was filled with stones to weigh it down. Folded linen was wound round thick wooden rollers which were then inserted under the box. The crank handle was turned and the box slowly trundled backwards and forwards over the linen. Such a machine was still in use in the estate laundry at Shugborough, Staffordshire in the 1920s. A handyman came every Wednesday to help the laundrymaids operate it. The firm of Bradford was the premier mangle manufacturer, and marketed the design well into this century.

The other design, typified by Tyndall's Scotch Mangle of 1850, recognized that box mangling was a grossly heavy job and that smaller items needed lighter treatment. Although bearing a strong visual resemblance to later wringers, it had three rollers. The principle was still to feed the dry linen through the first pair and back through the second.

Smaller mangles, or wringers, used simply for squeezing water from wet washing were developed during the 19th century. Five such machines were shown at the Great Exhibition. The process was cumbersome and more laborious than wringing clothes by hand, yet improvements by such firms as Ewbank made the wringer a more attractive proposition than a

As early as the 1850s the American firm of Lithgow advertised 'a gas-heated smoothing iron for tailors, hatters and family use'. The illustration shows a lady using the iron connected to the gasolier by a gutta-percha tube; the caption reads: 'Time alone must bring it into universal use.'

Ironing Simplified and Systematized.
100 PER CENT. SAVED IN LABOR AND FUEL.
LITHGOW'S PATENT GAS-HEATING SMOOTHING IRON, FOR TAILORS, HATTERS, AND FAMILY USE.

washing machine. Gearing systems became more sophisticated and weights for adjusting the roller pressure were superseded by springs with adjusting screws. Rollers of vulcanized rubber instead of wood caused less damage to fastenings. Wringers came with their own stands, or could be bolted on to a table, or – most conveniently – were an integral part of the washing machine itself.

Though larger items continued to be mangled, smaller ones were pressed smooth with heated irons, a process attributable to that well-known laundering nation, the Chinese. Ironing in Europe – and hence in the United States – was not common until the latter part of the 16th century. Designs of iron changed little before 1851, when three basic types were available. The box iron was simply, as its name suggests, a hollow and often ornate metal box into which was placed a previously-heated slug of iron. A companion slug would heat up while the first was in use. Box irons were both cleaner and presented fewer fume problems than the alternatives.

Charcoal irons were box irons into which glowing coals were placed. A pair of bellows was used to blast air on to the coals and so raise the temperature of the iron. There was a chimney to carry away the fumes. Using the charcoal iron needed some skill if the linen was not to be showered with disfiguring soot and glowing ashes.

The most popular model, and one which did at least avoid the languorous effect of fumes, was the 'sad' or flat iron. Because they were heated directly on the stove or fire, these were of limited cleanliness. They came in a variety of sizes but all were heavy to lift and hot to hold. A patent was taken out by J Whitehouse in 1865 for detachable and wooden handles which could be fitted to the iron in use.

From the 1860s, larger establishments might make use of one of Smith & Wellstood's new laundry stoves which had the fire enclosed so that the base of the iron didn't get dirty on the coals. Although flat irons were to continue in use even after the Second World War, it was clearly an attractive proposition for inventors to set to work on a self-heating and potentially cleaner appliance.

In 1861 the *Practical Mechanics' Journal* described an oil-heated iron, but this did not come into general use until later in the century when cheaper kerosene was

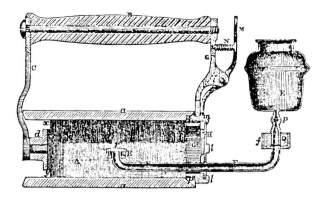

The Janus iron was heated by burning oil.

more widely available. Gas was potentially better suited to the task and, as most town houses by the end of the 19th century were being piped for gas lighting, it was more readily available.

Most people stuck with their flat irons, though some were prepared to use a gas burner to heat theirs. Later irons, heated by a continuous supply of gas, still had nothing positive to recommend them. As with oil irons, the fumes given off in a confined space must have been unpleasant to say the least and, as neither had any thermostatic control, it was probably easier to stick with a sad iron which did at least cool down. Not until the 1920s were the drawbacks of gas irons overcome but, by then, electric irons were beginning to make their mark.

In 1882 a patent was granted to H W Seely of New Jersey for a carbon-arc iron – but the idea was not a success. In 1895 advertisements appeared in Crompton's catalogues for an electrically-heated iron but, given that it weighed 14lbs (6.4kg), devotees would have had to have been strong indeed! Commercial products were available in America from about 1904 but, until the arrival of thermostatic control in the 1920s, there was little positive attraction, despite strong advertising. A well-known series of advertisements from the Westinghouse Company appealed to men's compassion: 'Put a stove in your office. You think it's hot, do you? Then what does your wife think while ironing to the accompaniment of a hot stove!'

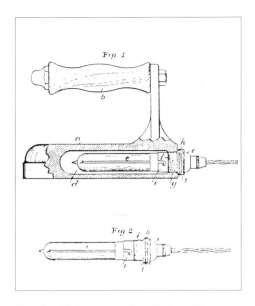

Rippingille's patent electric iron (Patent No. 13063 of 1913) was heated by lamps.

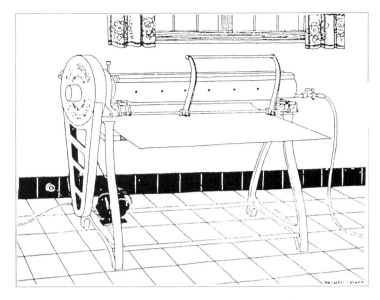

Thor rotary iron. Quick to realize the inefficiency of tackling large tablecloths, for example, with small irons, the Americans revamped the old dry roller mangle and gave it an electrical heating element.

Crompton's electric iron – complete with wooden plug casing.

A 'perfect gift' – the Morphy-Richards iron, 1939.

Sewing and sweeping

The link between these two otherwise disparate items is that both were early and highly successful subjects of American door-to-door sales techniques.

Sewing Machines

The initial idea of the sewing machine, as with the washing machine, was somehow to imitate the human action: in this case the movement of the seamstress's or tailor's needle. Early machine needles were forked or curved and pierced the fabric almost horizontally, but it was soon clear that the needle could not pass completely through the fabric and that the thread it carried must somehow be caught on the underside of the fabric. It was also clear early on that the eye of the needle must be at its pointed end.

One of the first ways of locking the thread was with a 'hooker' or 'looper', which caught a loop of thread on the incoming needle. The needle withdrew and, as it entered the next time, the thread passed through the previous loop and a new loop formed. The result was a single stitch on the top with a chain stitch underneath. This stitch – sometimes seen today for closing paper sacks and the like – has an inherent disadvantage in that if one end is pulled all the previous stitches will come out.

A solution was somehow to employ two threads, one on either side of the fabric, which would lock each other. Now, as the entering needle and thread left a loop, a boat-shaped shuttle carrying a bobbin wound with the underthread passed through that loop, the needle retracted, and the thread locked as the shuttle moved back ready for the next loop. In some designs, the bobbin was round and made a circular movement through the loop.

Needles were still curved and sometimes grooved to help catch the locking thread, and the fabric was held vertically by the pins on a plate. However, the fabric had to be reset once the stitching had moved the length of the plate. What was needed was a mechanism which

Elias Howe's sewing machine, about 1845. The cloth is carried past the needle vertically on the spiked feed-plate. The eye-pointed needle pierces the cloth in the direction of the handwheel.

Four stages of the rotating looper forming a chain stitch.

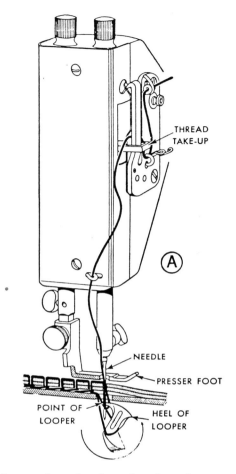

THREAD TAKE-UP

Ⓐ

NEEDLE

PRESSER FOOT

POINT OF LOOPER

HEEL OF LOOPER

1 Loop enlarged and previous loop drawn up.

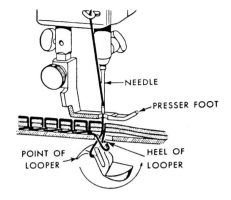

NEEDLE

PRESSER FOOT

POINT OF LOOPER

HEEL OF LOOPER

2 Loop further enlarged and previous loop completed.

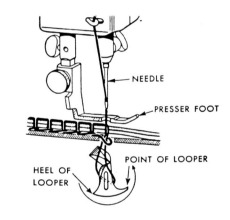

NEEDLE

PRESSER FOOT

HEEL OF LOOPER

POINT OF LOOPER

3 Looper entering loop of needle thread.

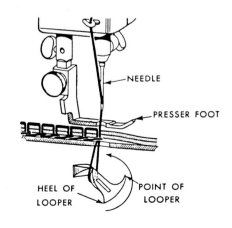

NEEDLE

PRESSER FOOT

HEEL OF LOOPER

POINT OF LOOPER

4 New loop being carried through previous loop.

both held the fabric and moved it through the path of the needle and bobbin at a steady rate.

Isaac Merritt Singer's contribution to the art was to gather together the ingredients of sewing machine wisdom and solve such teasing problems. He had seen a rotary machine being repaired in a Boston shop; unimpressed by the design he set about improving it. His subsequent design is much more recognizable to us than preceding ones. The needle passed at 90 into the fabric, which was held horizontally by a presser foot, and fed through the needle's path by a finely grooved vertical wheel moving in conjunction with the 'foot' which keeps the fabric pressed down. The shuttle

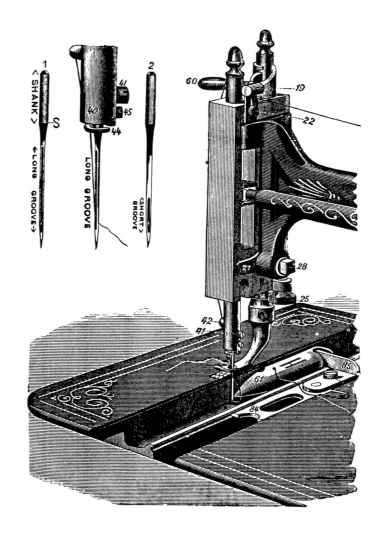

Elias Howe's improved lockstitch model.
The shuttle can be seen about to nose
through the loop of thread carried by the
needle.

moved forwards and backwards though in fact the stitch was locked in only one direction.

The machine, designed primarily for factory use, was operated by a treadle, leaving the hands free to control the work. Singer's machine was completed in September 1850.

The perpendicular element and presser foot had been used some 20 years previously in France by Barthélemy Thimmonier. His machines were used for making army uniforms but were smashed by disgruntled tailors anxious to preserve control over their livelihood. Nonetheless, Thimmonier went on to greater things and formed the first French sewing machine company.

The lock stitch had been developed in the United

States by Walter Hunt in 1834. Hunt, frustrated at being unable to solve the 'feed' problem, bowed out of the scene and sold his machine without patenting it. In fact, failure to patent, or infringements of designs that were patented, were a constant feature of sewing machine history – certainly in America. In 1846 Elias Howe patented his lock stitch machine in America; again, the feed mechanism was the ultimate frustration. Disillusioned, Elias sent his invention to London with his brother Amasa. There was little interest so Elias himself came to adapt his machine to sew corsets. When he returned home three years later, he found his American patent being infringed and other manufacturers making the profits Howe felt rightfully his. The matter went to court and Howe was eventually

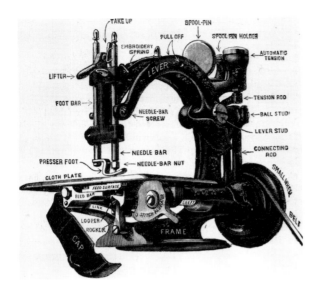

Willcox & Gibbs chain-stitch machine. The principle of the looper is explained on page 44. The Willcox & Gibbs instruction manual is at pains to point out that this chain-stitch machine (in their opinion) 'produces the only true lock-stitch.' It was just a matter of finishing the chain off properly by making sure that the end was caught through the last loop.

The Singer Manufacturing Company's factory at Kilbowie, near Glasgow, Scotland, 1867.

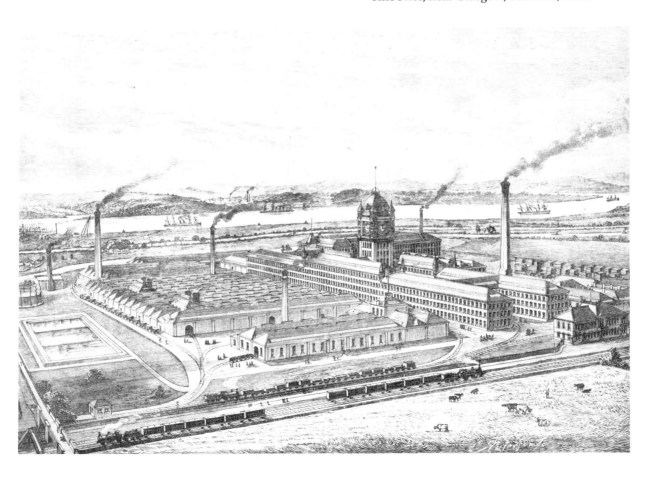

victorious. Singer too was found to have infringed Howe's patent; the solution to the barrage of claim and counterclaim was to form the Sewing Machine Combination (1856) whereby all major parties pooled their patents until the expiry of the last one in 1877.

Singer's flamboyant commercial instinct made his name synonymous with the sewing machine (rather like Hoover and the vacuum cleaner). Machines were sold not only door to door, but also through mail order firms. Payment could be made by instalments, and so made the purchase even more attractive. Later in the 19th century they were introduced into schools, especially in England, to capture the customers of the future. At a time when the female profile was very low where machines were concerned, Singer employed attractive women to demonstrate his machines in his Broadway shop window. He set up agencies all over the United States and always had one female demonstrator along with the mechanic and salesman.

Singer may have the fame, but another American probably contributed more than anyone else to sewing machine development – Allen B Wilson. With Nathaniel Wheeler, he formed the Wilson & Wheeler Manufacturing Company which produced more sewing machines in the 1850s and 1860s than any other company. Wilson's shuttle formed stitches on the backstroke as well as the forestroke. In 1854 he patented a four-motion feed, an idea used almost universally today. Serrated teeth catch the fabric, advance it, and return to repeat the motion, thus feeding it through at an even rate.

The sewing machine was not developed solely in the United States, though nowhere else matched the American inventiveness of the 1840s and 1850s. There

Wheeler & Wilson sewing machine on treadle stand. All the basic features of the stitch making mechanism are in place.

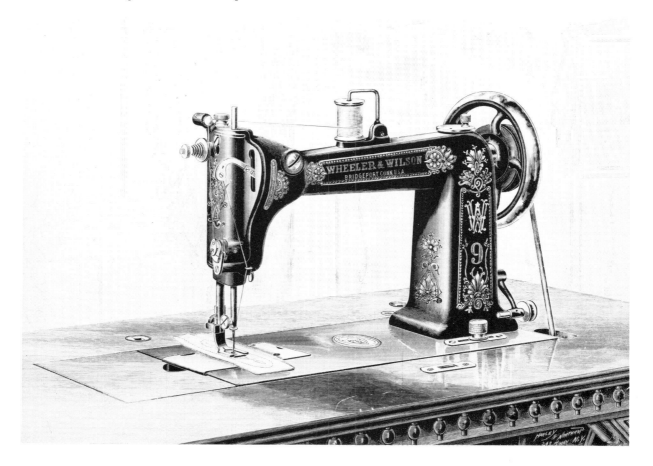

was a much slower start in Europe and only one British machine was shown at the 1851 Exhibition, an industrial lock stitch model by Charles Judkin. Pfaff and Frister & Rossman began manufacturing in Germany in the 1860s and Husqvarna in Sweden in the 1870s. The industry grew apace in Britain during this time with such companies as Wright & Mann of Ipswich, William Jones of Manchester and Bradbury & Company of Oldham. But the initiative remained American, with companies opening up agencies and factories in Britain: Wilson & Wheeler were in London in 1859, and Singer opened his first overseas factory in Glasgow in 1867. To complement the domestic sewing machine scene, Ebenezer Butterick of Massachusetts founded the Butterick Pattern Company in 1863 to produce paper patterns for home dressmakers.

Machines could be operated either by treadle or by hand. The former method, though efficient, was too redolent of the factory, and deliberately ornate, hand-operated machines were therefore developed in order to exploit the mainly-feminine domestic market. In fact, hand-operated machines often had to have heavy marble bases to prevent them moving during use.

By 1905, Singer had taken over the Wilson & Wheeler Company and was undoubted market leader, offering an unparalleled after-sales and part-exchange service. Other companies copied the Singer machine style and appearance. A Singer electric machine – complete with batteries – was available from 1889, though J H Cazal and Company of Paris had shown one at the 1867 French Exhibition.

Electricity was not the only source of energy; in 1880 Scientific American showed a Tyson motor – powered by gas, oil, steam or petrol – driving a sewing machine. Neat domestic steam engines, using a two-pint boiler, were also used; water-power, using a jet from the domestic supply was available; even animals in treadmills were suggested. But of course, as with other domestic appliances, none was a match for the clean efficiency of the small electric motor. Though the electric sewing machine began to come into its own after the First World War, hand-operated models retained their popularity in Britain for at least another 30 years – indeed, many thousands of antique but serviceable machines are kept for occasional use today.

Carpet Sweepers.

The Bissell Carpet Sweepers—
Bissell's Miniature, a ¾-size sweeper, suitable for ladies' use, and works conveniently among furniture 10/6
Bissell's Standard, a reliable Sweeper for all the lighter classes of work 10/6
Bissell's Grand Rapids, a '' Cyco-bearing '' pattern with greater brushing power ... 13/3
Bissell's Cosmopolitan, a '' Cyco-bearing '' pattern of the same power and type as the Grand Rapids, but more highly finished ; recommended for ladies' use 14/9
Bissell's Parlour Queen, a very powerful pattern for the thickest piles 17/3
Extra brushes for do. 2/5
  ,, bands, 0/7 ; wheel rings 0/3½
Toy Carpet Sweepers, Bissell's, The Baby, 0/9½ ; Little Jewel 2/8

Carpet Sweepers—
The '' Ewbank ''
(Royal),
12/6
Brushes (extra)
for do.
2/1

Two carpet sweepers with rotary brushes – Bissell's *Grand Rapids* **and the Ewbank** *Royal.*

Right
Booth's vacuum cleaner at work. The machine made so much noise that Booth was often sued for frightening passing horses. Eventually, he took a test case to appeal, and the Lord Chief Justice upheld his right to operate the machine in the street.

Carpet sweeping

Cleanliness is next to Godliness. As a reminiscer in Carol Adams' *Ordinary Lives* explained:

You sprinkled your tea leaves over the mat and they absorbed the bit of dust – then it's easier to sweep up. But you weren't considered a good house-wife unless you took your mats down to the yard, put them on the washing line and beat the life out of them.

Most people had rugs at home made out of rags. Rugs and carpets had to be brushed by hand and, as we can see from the above, Godliness was something to which even the working classes aspired. It was easier if you were better off, for you would have a housemaid – or two. But no matter how energetically the housemaid brushed and beat the carpets and curtains, the dust was merely raised – to settle elsewhere.

A street-sweeping machine had been developed by Joseph Whitworth in the 1840s and his idea of brushes revolving on a drum was the inspiration for domestic carpet sweepers. In 1876, Melville Reuben Bissell patented the first carpet sweeper and called it Grand Rapids after his home town in Michigan. Apparently Mr Bissell, a china-shop owner, was allergic to the dust which accompanied the straw in which his china was packed and sought some mechanical device to alleviate his problem.

His device was an outright success and, though various other companies – notably Ewbank – both in the United States and in Britain marketed their own models, the name of Bissell was dominant. In the 1880s Bissell machines were used 'daily in the households of HM The Queen and HRH The Princess of Wales'. There were Babies (for children), Miniatures (for ladies), Standards, and Parlour Queens, 'a very powerful pattern for the thickest piles'. In the 1930s aluminium castings replaced wooden parts, making the machines much lighter and easier to use.

The carpet sweeper's rival was much more efficient.

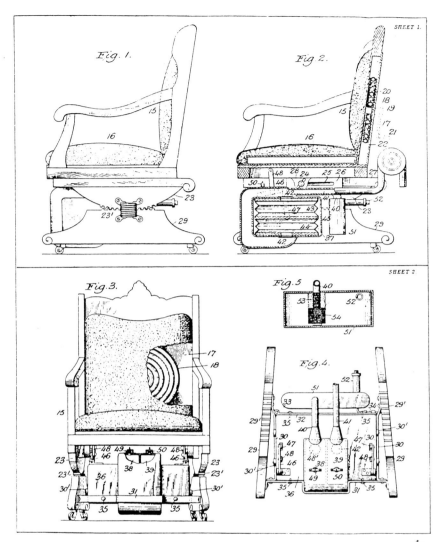

Fig. 1.

Fig 2.

Fig. 3.

Fig. 5.

Fig. 4.

Behringer's invention (Patent No. 11413 of 1908) uses the action of a rocking chair to provide the suction.

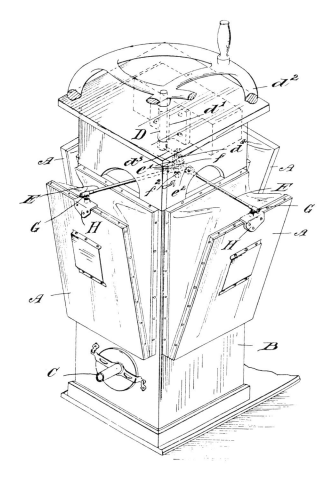

Davies' device (Patent No. 26676 of 1909) has four
bellows operated in sequence as the handle is turned

In 1901, Hubert Cecil Booth invented the first successful vacuum cleaner. Cleaning by suction had first been
explored in the United States in 1859, largely for factory cleaners, but the success was so minimal that the
idea lay dormant until 1901, when Booth was inspired
by yet another failure. At the Empire Music Hall, London, he witnessed a demonstration of an American
cleaner which blew compressed air through the carpets: the result was a cloud of dust which settled again,
restoring the status quo. Booth's immediate reaction
was to experiment again with sucking the air; what
made his new machine successful, rather than the
thirty-odd inventions on similar lines in the last three
decades of the century, was an effective source of
power.

Booth's first machine, Puffing Billy, had a 5 hp piston
pump driven by a petrol – gasoline – engine or an
electric motor. There were no brushes; dust was ex-
tracted entirely by powerful suction through nozzles
at the ends of long flexible tubes.

Too large for domestic application, Booth's machines would be permanently installed in larger buildings, with the tubes fitted to sockets in each room,
rather like electric sockets. More commonly, the bright
red mobile units of the British Vacuum Cleaner Company would draw up outside the building. Yards and
yards of hose would pass through the windows to
reach and clean all parts inside. The most famous
BVCC engagement, and the one which made its name,
was to clean the carpets in Westminster Abbey for the
coronation of Edward VII and Queen Alexandra in
1901. This led to many prestigious demonstrations,
including one at Buckingham Palace. Their Majesties
were astonished at how dirty the place was and
promptly ordered a machine.

Fashionable society held vacuuming tea parties to
watch the cleaning take place; part of the tube was
made of glass so that spectators could see the dust
vanishing into the machine.

Booth was in fact singularly reluctant to sell his machines, preferring to hire them out. On the other hand,
the equivalent machine in the United States – first installed by David T Kenney in the Flick Building in New
York in 1902 – was for sale, and these machines
achieved considerable popularity in America.

It was not clear at this stage, however, that the portable vacuum cleaner was the machine of the future.

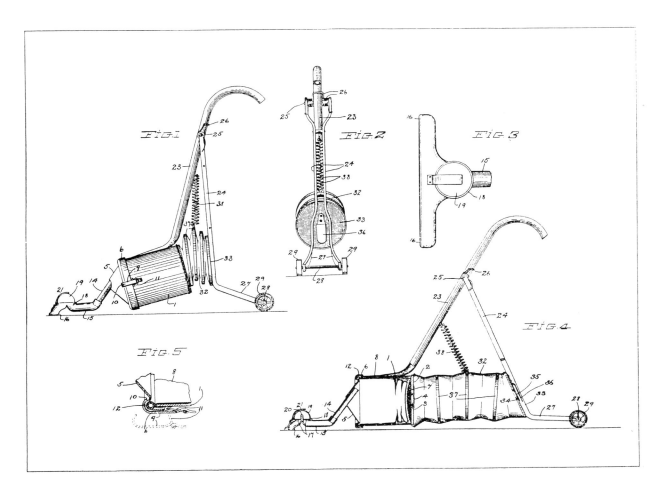

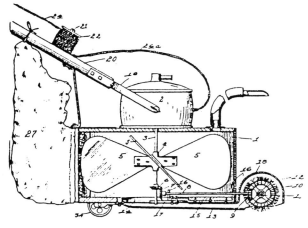

Kirby's design (Patent No. 6055 of 1912) requires the machine to be advanced in the manner of a caterpillar.

Spangler's patent drawing: filed in the United States 14 September 1907; in Britain, it was Patent No. 18101 of 1908.

The Electrolux (1926) 'is so simple that even a child can use it. One of its outstanding features is that it disinfects as it cleans.'

Many and varied were the initiatives for creating the necessary mechanical suction. Booth's smaller version for retailing, the Trolley Vac of 1906, weighed 100lb (45kg) and was very expensive. For the sort of household where appearances mattered, enough servants would have been available to operate the two-man cleaners such as the Griffith (1906) or that patented by D E Davies in 1909 in which a rotating wheel operated four bellows. A reduced version of this became the Wizard of 1912.

For the single operator, methods were even more ingenious. A patent filed in 1908 by E & H Behringer relates to the rocking motion of a chair operating the apparatus. The operator had an especially long hose, as did the operator of K von Meyenburg's back-pack bellows. Other one-man cleaners had the nozzle on the machine itself, so various hand pumps were em-ployed. The most popular of this type was the Star. Resembling an enormous bicycle pump in reverse, it was cheaper than electrically powered machines, was fairly efficient, and survived on both sides of the Atlantic into the 1930s in homes where electricity had not been installed.

Although a British version was patented less than a year later – one which could moreover deal with a 'variable nap' – it fell to an American to make the critical leap forward.

In 1907 an asthmatic janitor who wanted to make his job easier – one James M Spangler of Canton, OH – constructed a machine from tin, wood, a broom handle and a pillow case. An electric motor was coupled to a fan disc and a cylindrical brush. This portable vacuum cleaner, though crude, worked excellently. Spangler was related to W H 'Boss' Hoover, a manufacturer of

Madame, if you had X-RAY eyes —

If you had X-ray eyes, how simple shopping would be; how much money you would save; what disappointments you would avoid! For instance, this year several hundred thousands of women will buy vacuum cleaners. If they had X-ray eyes, all would choose the HOOVER. They would see that in the HOOVER they could buy 3 cleaners in one for the price of one! They would see too, how much more thoroughly the Hoover cleaned, how its patented Agitator tapped loose and vibrated to the surface the deeply buried dirt that ordinary suction cleaners cannot remove.

As you cannot acquire X-ray eyes, do the next best thing and post the coupon below for booklet describing the various Hoover models.

YOUR OLD CLEANER IN PART EXCHANGE

THE DE LUXE
ONLY £1 DOWN

The HOOVER
It BEATS ... as it Sweeps ... as it Cleans

COUPON TO HOOVER LIMITED, Dept. I.H.3 PERIVALE, GREENFORD, MIDDLESEX.
Please send me particulars of all Hoover models, and easy payments.

NAME...

ADDRESS ....
....

THE HOOVER
FAMILY RANGE
Dustette ... £4.19.6
*Junior £10.15.0
*Popular... ...£17. 5.0
De Luxe With-
out Cleaning
Kit £19.19.0
Ensemble De
Luxe ... ... 22 gns.
*Cleaning Tools extra
according to model.*

The Hoover (1939) – 'the next best thing to X-ray eyes.'

Tellus in Sweden had a particularly adaptable version capable of drying clothes or hair, working as a room heater, or as a cooking fan or even spraying paint (1931).

leather goods. 'Boss' wanted to diversify his business to begin making and marketing the cleaners. 'Spanglering' didn't have a chance to catch on, but 'Hoovering' did. Hoovers were excellent for selling door to door, and a large retail organisation was quickly built up. Machines were being exported to Europe by 1913, but their reception there was not quite as euphoric. In Britain at least electrical power was much slower to become available; there was also a good deal of suspicion about electrical machines generally and some doubt as to whether the servants could cope with them.

However, not to be outdone, rival companies in Britain quickly developed their own alternatives (the Goblin, the Magnet and Universal). In 1926 Hoover added a beater rod to the cylinder so, still beating as it swept as it cleaned, the job was made a lot easier than it had been in the days of sprinkling tea leaves.

Gadgetry

Food preparation

We illustrate ... a great variety of labour-saving devices which tend to make housework a pleasure instead of drudgery.

So proclaimed the Sears Roebuck catalogue of 1902 at the start of its hardware section. Given the repetitive and time-consuming activities entailed in preparing food for large households, cooks and kitchen maids alike were disposed to welcome any machine which made their job easier. Nowadays, even small households have choppers, mincers and processors. With easy-to-clean and easy-to-assemble parts, tasks of food preparation are indeed less time-consuming even though the degree of pleasure involved may be debatable. One cannot escape the feeling, however, that the orgy of gadget designing which began in the 1870s owed more to the satisfying of their inventors' creative urges than the meeting of any stated need.

Chopping, grinding and mincing machines have had the most enduring popularity in kitchens. Among the earliest manufacturers were Nye and Lyon in Britain and Enterprise in America. From the mid-1850s, these manufacturers were producing sausage machines, food masticators (to assist digestion, loss of teeth, etc), vegetable and herb choppers and slicers. The largest of Lyon's mincers had a flywheel and could cut 57 lb (26kg) of meat in 25 minutes. This is not surprising; most large-scale mincers, for use in the preparation of animal feed for example, were highly efficient. The problem was to reduce the size without losing the efficiency.

In the 1880s, Spong and Co of London came to dominate the market. Spong's one mincer could, with various attachments, do the jobs of several individual machines – all using a screw principle to drive the food forward. The alternative principle was a gearing system, developed by David Lyman of Connecticut in 1868, and adopted by the American Meat Chopper in

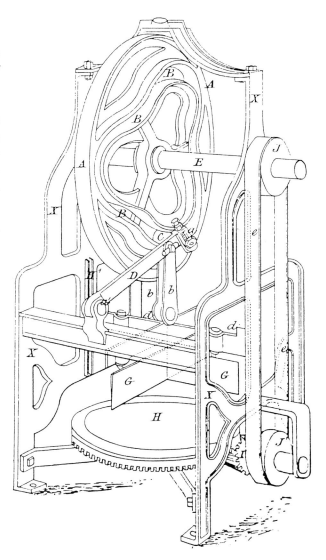

Black's patent (No. 1933 of 1858) chops up the food. The inventor suggested that it might be driven by a water-wheel.

1879. It was an interesting machine in that the 'tub or receptacle used receives an intermittent or continuous rotary motion'. 'I've just invested in a mincing machine and cannot think how I ever managed without it', commented the fictional Margaret Trent in the *Girl's Own Paper* of March 1882.

Other time-consuming jobs related to the preparation of fruit and vegetables. Potato and apple peelers were popular in America; less so in Britain, where cooks couldn't abide the waste that inevitably resulted from apples and potatoes that lacked uniformity. Peas and beans had to be shelled and small fruit such as raisins and cherries pitted, and Mr Smith's machine of 1865 could do just that. Removing the pits (stones) from raisins is one of the strands of the sub-plot in the drama of kitchen gadgetry as it emerges from the patent books. All sorts of designs were thought up to alleviate what clearly must then have been a pressing task; how fashions have simplified!

Egg-boilers and openers also seem to have posed teasingly interesting problems for late-Victorian inventors. A really exciting apparatus patented by J Durrant in 1864 automatically removed the eggs from the water when boiled: 'Their ends are cracked by being brought against the top of the vessel, a flap falls down to allow escape of steam and a bell is rung.'

An illustration of the way in which fashion and inventiveness came together is to be found in the domestic ice-cream makers which achieved a brief vogue in Britain towards the end of the 19th century. One of the leading names in Victorian ice circles was H C Ash. He had been making and designing chests and safes for years, and in 1860 developed a piston freezing machine. This machine consisted of two buckets; a handle on the lid rotated a paddle or worked the piston in the inner bucket, a process which took about 20 minutes of hard work. Twenty years later Household Management observed:

Ice is now so much in use at English tables that it has become a necessity of household economy and dessert ices follow summer dinners as a matter of course. Dessert ices are by modern invention placed within the reach of most housekeepers and it is a pleasant and easy amusement for ladies to make ices by Mr Ash's Patent Piston Freezing machine.

A suitable comment on the time must be the fact that, when the small electric motor finally made its way into food preparation, it was in the important areas of

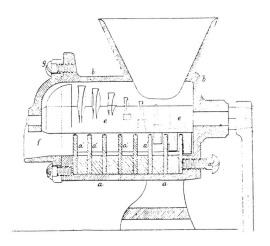

Turner's patent (No. 1869 of 1862) shears the food between fixed and rotating blades.

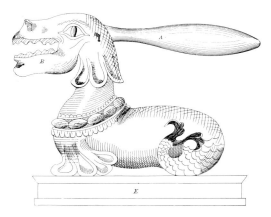

W E Gedge's patent masticator (No. 1191 of 1874) 'assists digestion, loss of teeth etc'.

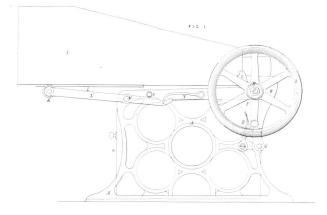

E G Smith's patent machine (No. 1244 of 1865) for shelling peas and beans and stoning [pitting] small fruit.

cutting, choppping and mincing. The Universal, an electric mixer, was available in America from 1918, but was less in demand in Britain because of the decline in the numbers of large establishments and the slow spread of electricity. Hand-turned mincers, beaters and blenders remained popular until well after the Second World War, though with parts mercifully easier to dismantle and clean, and without the dubious metal finishes of their forebears.

Before the introduction of stainless steel in the 1920s, cleaning all kitchen utensils was demanding, but cleaning knives was particularly arduous. The classic knife cleaner to replace the traditional knife board was William Kent's Rotary Knife Cleaner of 1882. After being washed, the knives (only four in the 1882 version) were inserted into the slots around its rim, and a

A M Clarke's patent egg opener (No. 1576 of 1869).

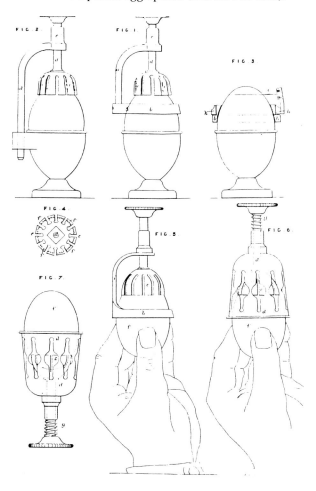

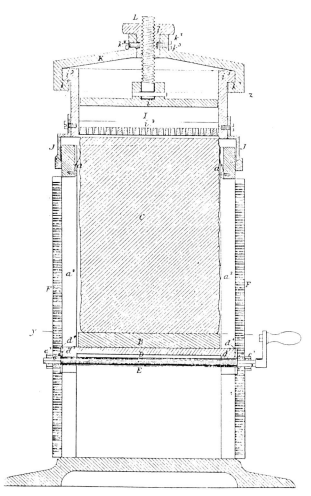

W Berry's patent butter spreader (No. 792 of 1865). What but the huge quantities of neat sandwiches to be made for tea-time callers can have prompted the search for an efficient butter-spreader? And how long did it take to clean?

special abrasive powder poured into another hole. The operator then turned the handle, and felt pads and bristles scoured over the blades which emerged bright and shining after a dozen or so turns. The machine also sharpened the knives – provided they were inserted the right way!

Dishwashers

Dishwashers feature prominently in mid-century patent books and trade journals but, as with early wash-

UP-TO-DATE HOUSEHOLD APPLIANCES

Designed specially for the smaller kitchen.

by
MARY WHIRTER.

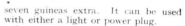

*T*HIS new electrical appliance for cookery is fully described in the article. On the left it is shown freezing ice cream, on the right sifting flour and sugar, and below cutting up vegetables with the aid of the food chopper attachment. It can be connected to any light socket.

*T*HE kitchen is sometimes curiously neglected when the electric power and light points are planned in new houses. In the modern kitchen there should be at least two—one near the sink for the washing machine and electric iron, and the other within easy reach of the kitchen table for electric cooking apparatus.

Although there are fewer surfaces to catch and hold the dust than in other parts of the house, the vacuum cleaner will often be found most useful in the kitchen for blowing dust from dresser shelves, burners of the gas stove, and corners which are difficult to reach with mop or brush in the ordinary way.

For these a small hand vacuum cleaner will be found a useful auxiliary to its bigger brother; it is light, has no attachments, and the suction can be converted to a blowing action by a mere turn of the switch. It is particularly good for such things as mattresses, cushions, upholstery and stair carpets.

Among the larger cleaners, a new electric motor brush driven model has just been put on the market in this country. It does not clean by suction only. The rapidly revolving brush vibrates the dirt from the carpets, and the powerful suction then draws it into the bag.

It would be most suitable for thickly carpeted floors, and, as it has no accessories, a hand vacuum cleaner as described above is indispensable for use with it.

For the electrical preparation of food, you can confine your appliances to one or two machines, each for a specific purpose, such as an electric egg and batter mixer and beater, a fruit juice extractor, and perhaps a coffee roaster, this last a recently invented appliance which brings the joy of freshly roasted and ground coffee within the reach of everyone.

*T*HERE is, however, a new electrical appliance which has recently made its appearance, and which automatically deals with practically every aspect of the preparation of food.

Its size is not its least recommendation, for it stands but eighteen inches high, and takes up no more room than a corner of the kitchen table. The machine alone costs £24, the cabinet with white porcelain enamel top being

seven guineas extra. It can be used with either a light or power plug.

It saves both time and labour in pastry and bread making, for it will mix flour and milk, beat eggs, knead the dough, and sift flour and sugar, all of which processes take some time when done by hand.

In ordinary cooking it will mix meat and seasoning, and with the special food chopper attachment cut up meat and vegetables, nuts, raisins, dates and the like, and also mash potatoes.

For the housewife who has hitherto hesitated at buying a washing machine on account of cramped space or of the heavier household linen being sent to the laundry, there is now a miniature electric portable washer costing twelve guineas.

Some idea of its capacity can be gleaned from the fact that it will take five men's shirts, or a pair of cot sheets and one pillow case at one filling.

Washing, rinsing and drying are done in the one tub, a consideration in a small kitchen, and the current consumption is approximately that of a 100 c.p. electric bulb. Washing takes from ten to fifteen minutes, the clothes are then dried by centrifugal force, "mangle dried" in about three minutes, or ready for ironing in from ten to twenty.

It has been found that this washer is particularly kind to artificial silks and woollens, which emerge from it soft, fluffy and unshrunken. The machine can be put on the draining board.

When a wringer is used, it is best to get either the kind which converts into a table or one to clamp on to a folding stand. A compact table model I have in mind converts from table to wringer using only two fingers, and has extra resilient rubber rollers to squeeze out the last drop of water. It is well geared to reduce work to a minimum.

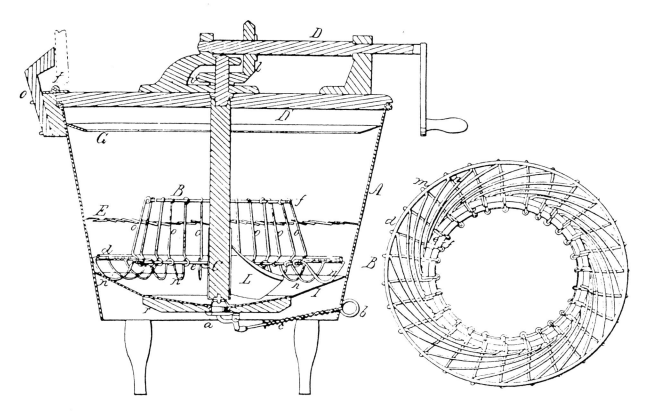

Above

LA Alexander's US Patent No. 51000 of 21 November 1865. Turning the handle whirled the cradle full of dirty dishes through the water.

Right

W J Schlesinger's patent machine (No. 1977 of 1870) was simply a water-bath lined with zinc, carrying an open tray or rack to contain the articles. A 'dasher' driven by a crank handle revolved around the tray, driving water between the articles.

Left

'An up-to-date household appliance designed specially for the smaller kitchen freezes ice-cream, sifts flour and sugar, and cuts up vegetables with the aid of the food chopper attachment. It can be connected to any light socket' (1931).

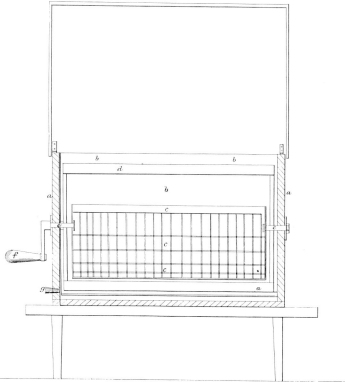

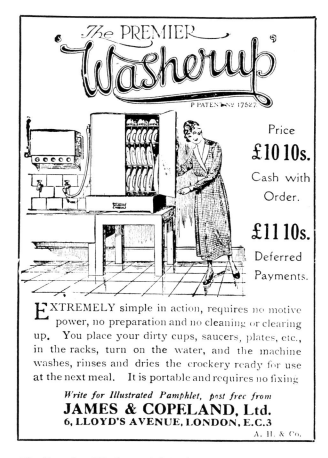

EXTREMELY simple in action, requires no motive power, no preparation and no cleaning or clearing up. You place your dirty cups, saucers, plates, etc., in the racks, turn on the water, and the machine washes, rinses and dries the crockery ready for use at the next meal. It is portable and requires no fixing

Write for Illustrated Pamphlet, post free from
JAMES & COPELAND, Ltd.
6, LLOYD'S AVENUE, LONDON, E.C.3

A. H. & Co.

The Premier 'Washerup', (1920).

ing machines, they were geared to large commercial establishments where water was heated and a form of power (usually steam) was available – for example Daguin's machine of 1855.

C E Hope-Vere's machine of 1875 forced jets of water on to the dirty articles. This was a popular method which appeared during a spate of dishwashing inventions in the early years of this century – for example A W Bodell's machine of 1906.

The alternative method was to use a pump to circulate the water between the articles, as in Goldman's machine of 1905. M Fridjian's design of the same year repeatedly plunged a perforated casing containing the dishes into a tank of water.

In most people's eyes, the dishwasher was a supremely new-fangled gadget, and a general apathy inhibited its development. In 1923 *House and Garden*

ran an article entitled: 'Overcoming the drudgery of the Dishcloth'. This offered advice on the three basic types of dishwasher then available. One consisted of nothing more than: a stream of hot water from a washing nozzle attached to the hot water tap and directed by hand. The nozzle contains a soap mixer operated by a thumb lever, so that soapy water is delivered for washing and clear for rinsing. This is only suitable when the supply of hot water is plentiful.

Another, though this time an electric one, was also hand filled – from a kettle. What followed next is unclear. Presumably the electric motor drove plungers which sent the water over the dishes. Yet another, the Blick: is one of the simplest and best of machines worked by hand ... The plates stand in a special rack above the water which is thrown over them by a revolving propellor.

The Blick follows the same principle as Schlesinger's of 1870; however, it is pre-dated by an American patent of 1865, and differs from some machines today only in the level of sophistication.

For all that, early machines did not wash the dishes well enough, and it was not until improved soaps were developed in the 1950s and 1960s that the machines became efficient enough to warrant consumer attention.

Refrigeration

Apart from a commercial interest in refrigerated fresh food exported from Australia and South America, the British showed far less enthusiasm than the Americans for storing and making ice. Certainly ice chests or ice caves were exhibited in 1851 and continued to be sold. These needed a supply of ice which, of course, even with fairly effective charcoal insulation, eventually melted.

In 1882 P Jenson designed a 'refrigerator for domestic use' wherein cold water circulated around the food chamber. This was a favourite method but there was no real alternative to the use of ice for cooling, nor indeed for preserving food over a period of time. Ice was being commercially produced but by massive systems whose reduction in size posed the biggest problem to the development of the domestic refrigerator.

The action of a refrigerator depends on the absorption of the 'latent heat' of a liquid or liquefied gas when

The Barnet household refrigerator, 1921.

it evaporates. The action is cyclical and the refrigerant re-condenses after evaporation. In the ammonia refrigerator, the vapour is reabsorbed by water; a neater method compresses the vapour.

Almost inevitably, given their longer historical attachment to ice, the initiative was all American. In 1913 the first domestic refrigerator, the Domelre (*Dom*estic *El*ectric *Re*frigerator) went on sale in Chicago, rapidly followed by models from Kelvinator in 1916 and Frigidaire in 1917. In these early appliances, the mechanism was still extremely bulky and took up almost half the total space. The compression refrigerator was considered marginally more flexible despite the relatively silent operation of the absorption designs. The fundamental reason for one system prevailing over another was largely an historical accident. General Electric moved into the refrigerator market early in the 1920s, taking over Kelvinator and Frigidaire. Later, General Motors and Westinghouse also moved into

The Electrolux refrigerator – 'the only refrigerator in the world to operate continuously by electricity, gas or paraffin. It makes all the ice you need, and enables you to prepare delightful desserts, and to have a wide range of delicacies always available. *It keeps food fresh for an indefinite period.*'

the field; the absorption companies were simply unable to finance the development or promotion of their products.

In Britain, refrigerators were viewed as an unnecessary luxury. Frigidaire commented: 'The hard sell was probably essential in a Britain which regarded ice only as an inconvenience of winter-time and cool drinks as an American mistake.' The first British Frigidaire was sold in 1924 but not until after the Second World War did the sale of refrigerators take off in Britain.

Chronology

1596	Sir John Harington's water closet	1882	Jenson's domestic refrigerator
1767	Feetham's shower bath		William Kent's knife cleaner
1775	Cumming's water closet	1884	Sugg's *Domestic Uses of Coal Gas*
1830	Demonstration of cooking by gas	1887	Bostel's modern closet
1831	Faraday's transformer	1889	Commercial electric motor
1846	Howe's sewing machine	1891	Cooking by electricity
1850	Singer's sewing machine	1901	Booth's vacuum cleaner
1851	The Great Exhibition (London)	1906	Nichrome wire for elements
1859	Mrs Beeton's *Household Management*	1907	Spangler's vacuum cleaner
1860	Ash's piston freezing machine	1913	Stainless steel
	Revd Henry Moule's earth closet	1923	Regulo gas-oven thermostat
1868	Maughan's 'Geyser' gas water heater	1924	The Aga solid-fuel cooker
1870	Mann's siphonic closet	1926	The Electricity (Supply) Act
1876	Bissell's carpet sweeper		(England & Wales)

Further reading

PERIODICALS

The Expositor

House and Garden

The Housewife

Ideal Home

Ladies' Home Journal

Practical Mechanics' Journal

Scientific American

Sears Roebuck Catalogue 1902

BOOKS

Carol Adams
Ordinary Lives
Virago

Gareth Anderson
Machines at Home
Lutterworth Press 1969

Army & Navy Stores
Yesterday's Shopping: Reprint of 1907 catalogue
David and Charles

Isabella Mary Beeton
Household Management
Ward, Locke & Co Ltd 1861

Grace R Cooper
The Sewing Machine: its Invention and Development
Washington, DC: Smithsonian 1977

Ruth Schwartz Cowan
More Work for Mother
Basic Books, NY 1983

Caroline Davidson
A Woman's Work is Never Done
Chatto and Windus 1968

David J Eveleigh
Firegrates and Kitchen Ranges
Shire Publications 1983

Stuart Galishoff
Pollution and Reform in American Cities 1870–1950
Texas 1980

S Giedion
Mechanisation Takes Command
OUP New York 1983

George & Weedon Grossmith
Diary of a Nobody
Many publishers 1892 to date

David de Haan
Antique Household Gadgets and Appliances
Blandford, Poole 1977

HJ Habakkuk
American and British Technology in the 19th Century
CUP 1962

Christina Hardyment
From Mangle to Microwave
Polity Press 1988

Molly Harrison
Home Inventions
Usborne 1975

Carol Head
Old Sewing Machines
Shire Publications 1982

Barbara S Janssen ed
Icons of Invention: American Patent Models
Washington, DC: National Museum of American History 1990

Gertrude Jekyll
Old English Household Life
Batsford 1925

Glenna Matthews
Just a Housewife: Rise and Fall of Domesticity in America
OUP New York 1987

L J Peek & L E Slater
Household Equipment
John Wiley & Sons New York 1950

Pamela Sambrook
Laundrey Bygones
Shier Publications 1983

Winifred D Wandersee
Women's Work and Family Values
Harvard 1981

Lawrence Wright
Clean and Decent
Routledge and Kegan Paul 1960

Laurence Wright
Warm and Snug: The History of the Bed
Routledge and Kegan Paul 1962

Lawrence Wright
Home Fires Burning: The History of Domestic Heating and Cooking
Routledge and Kegan Paul 1964

D Yardwood
Science in the Home
Batsford 1983

Index

The PREMIER Washerup

P PATENT Nº 17527.

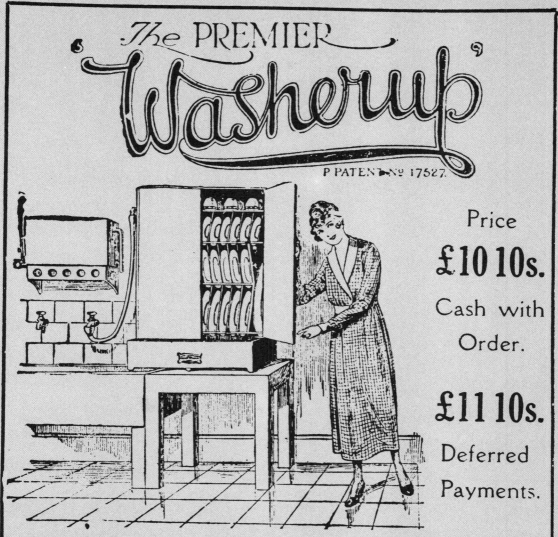

Price
£10 10s.
Cash with
Order.

£11 10s.
Deferred
Payments.

EXTREMELY simple in action, requires no motive power, no preparation and no cleaning or clearing up. You place your dirty cups, saucers, plates, etc., in the racks, turn on the water, and the machine washes, rinses and dries the crockery ready for use at the next meal. It is portable and requires no fixing

Write for Illustrated Pamphlet, post free from
JAMES & COPELAND, Ltd.
6, LLOYD'S AVENUE, LONDON, E.C.3

A. H. & Co.